SUSTAINABILITY FOR HEALTHCARE MANAGEMENT

Sustainability is not unique to health, yet sustainability is a unique vehicle for promoting healthy values. This book challenges healthcare leaders to think through the implications of our decisions from fiscal, societal and environmental perspectives. It links health values with sustainability drivers in order to enlighten leadership about the value of sustainability as we move toward a new paradigm of health.

Fully updated for the second edition, the book now includes case studies about:

- Waste disposal and cost
- Chemicals of concern
- Cost of water
- Green building ratings

This book is a unique resource for researchers, students and professionals working in health and healthcare management because the book connects key concepts of environmental sustainability with healthcare operations. Readers will gain an appreciation for translating leadership priorities into sustainability tactics with beneficial operational outcomes.

Carrie Rich is the co-founder and CEO of The Global Good Fund and serves on the adjunct faculty at George Washington University, USA and on the faculty at the Amani Institute, Internationally. She is the 2016 EY Entrepreneur of the Year, and recipient of the POLITICO Women Who Rule Award.

J. Knox Singleton is CEO of Inova Health System, a leading not-for-profit healthcare system in the Washington DC region. He is co-founder of The Global Good Fund, and co-founder and past president of the Community Coalition for Haiti, which creates community-driven solutions in healthcare, education, and community development in Haiti.

Seema S. Wadhwa is the Assistant Vice President of Sustainability and Wellness for Inova Health System. Ms. Wadhwa has helped drive the link between sustainability and wellness in the healthcare industry while leading the Healthier Hospitals Initiative from launch to over 1,300 hospitals, through her work at Inova and while sharing her passion through education. Seema is most fulfilled when serving in the local and global community.

SUSTAINABILITY FOR HEALTHCARE MANAGEMENT

A Leadership Imperative

Second Edition

Carrie Rich, J. Knox Singleton and Seema S. Wadhwa

Routledge
Taylor & Francis Group

LONDON AND NEW YORK

from Routledge

Second edition published 2018
by Routledge
2 Park Square, Milton Park, Abingdon, Oxon OX14 4RN

and by Routledge
711 Third Avenue, New York, NY 10017

Routledge is an imprint of the Taylor & Francis Group, an informa business

© 2018 Carrie R. Rich, J. Knox Singleton and Seema S. Wadhwa

First edition published by Routledge 2013

British Library Cataloguing in Publication Data
A catalogue record for this book is available from the British Library

Library of Congress Cataloging in Publication Data
Names: Rich, Carrie R., author. | Singleton, J. Knox, author. | Wadhwa,
Seema S., author.
Title: Sustainability for healthcare management : a leadership imperative /
Carrie Rich, J. Knox Singleton, Seema S. Wadhwa.
Description: Second edition. | Milton Park, Abingdon, Oxon ; New York,
NY : Routledge, 2018. | Includes bibliographical references and index.
Identifiers: LCCN 2017047068 (print) | LCCN 2017047922 (ebook) |
ISBN 9781315276878 (eBook) | ISBN 9781138244511 (hbk) |
ISBN 9781138244528 (pbk) | ISBN 9781315276878 (ebk)
Subjects: LCSH: Medical care–Management–United States. | Medical
care–Administration–United States.
Classification: LCC RA971 (ebook) | LCC RA971 .R527 2018 (print) |
DDC 362.1068–dc23
LC record available at https://lccn.loc.gov/2017047068

ISBN: 978-1-138-24451-1 (hbk)
ISBN: 978-1-138-24452-8 (pbk)
ISBN: 978-1-315-27687-8 (ebk)

Typeset in Bembo
by Wearset Ltd, Boldon, Tyne and Wear

For our children

CONTENTS

FIGURES

TABLES

FOREWORD

sustainability
Pronunciation: /sə,stānə'bilitē/
noun

There is no alternative to investing in a sustainable future; healthcare and the communities we serve are damaged by the failure to achieve sustainability objectives. There is no doubt about the threat to fiscal, environmental and human health if we maintain the status quo. Healthcare is part of the problem.

In every community healthcare can be part of the solution. Healthcare's stake is unique. Although unable to independently battle the forces that contribute to the problem, healthcare can be a model of response at the local and societal levels, a source of expertise and a catalyst of action for a sustainable future.

Are the full implications of healthcare sustainability well known? Certainly not by all. Many healthcare organizations will be content with traditional roles of community benefit. Traditional community benefit is an immensely important contribution, but at a level of engagement that falls short of potential contribution and community responsibility. The achievement of sustainability requires a broader engagement by healthcare.

The role of the hospital is evolving as medicine advances both scientifically and organizationally, from being the curative center to being the center of health. The new reality is that the hospital institution has a unique responsibility to address the challenges of sustainability to the extent that the hospital is responsible for community health. Hospitals must reach for a new definition of the health of the community, one that goes beyond the traditional measures of community benefit. The hospital is the only institution that has the mission, intellectual resources, reach into centers of influence, ability to command attention, and leadership potential to help guide the community toward sustainability. That is the full implication of sustainability for the hospital and the basis of a new social compact.

Begin forward

The mission of this book is to catalyze transformational change in the healthcare industry through sustainability. This book is not about tree hugging, nor is it about affection for Mother Nature, though there is a time and place for that. There is more to sustainability than environmentalism. The environmental movement does, however, provide a platform for the emergence of sustainability as the new mantra that integrates economic, political and moral forces to address the threats to the environmental dimensions of well-being. The essence of the new compact is clear. Through sustainability, healthcare renews its commitments to prevention and wellness, strengthening its compact between provider and community. Converting this compact into explicit, achievable objectives that address sustainability implications for a given community is a long, complicated process. It requires informed, thoughtful and committed professional healthcare administrators to focus institutional attention and energy on the consequences for the hospital and the health of the community.

The challenge of sustainability mandates an expansion of the scope of health administration professionalism

What value does the healthcare organization add to the community? It is in the answer to this question that the role of healthcare sustainability is addressed. If the value of the healthcare organization to its community is restricted to curative care and serving as a "doctor's workshop," then the hospital has what may be defined as a limited value framework (the hospital performs quality medical care). The administrator with a limited value framework says, "We are not in the business of healing the environment. We take care of sick people." The weakness of this limited value framework is that it does not align with the community's health needs. Making people whole again is an important part of healthcare needs, but it is only one dimension of a community's broader health needs.

How does the broader value framework get built into the DNA of the hospital? There is at least one set of people with the perspective to look across resources and into the community, to envision a broader value impact upon the community, which we call a healthy community. Not every hospital or healthcare leader will be prepared for this challenge, but most hospital administrators are positioned to aggregate skills and resources as valuable contributors to achieving the community's healthcare sustainability goals. Executive leadership sets the tone.

If you visit an average town of 30,000 people and look around, you will find that a hospital is an indispensable community resource, woven into the fabric of community life. Every community depends on access to a hospital. The person who runs the hospital is probably ideally positioned (due to status, access to resources and place in the community) to improve the quality of life in terms of health and well-being, to create sustainable conditions in which people are most likely to flourish. Healthcare leadership has an opportunity to be the first citizen of sustain-ability because it is a dimension of healthcare. It can be a calling but above all it is a mandate: to be the person who raises the issue, facilitates the conversation and brings together the players. The hospital becomes a model, a demo for the sustain-able citizen from which the entire community benefits. That role requires visionary leadership to bring a set of values and a sense of values not only to, but for, the community. Sustainability is central to the concept and definition of healthcare, and is therefore the business of healthcare and healthcare leadership.

Gary Cohen, co-founder and executive director of Health Care
Without Harm and a 2015 MacArthur Fellow

PREFACE

There is no alternative to investing in a sustainable future; healthcare and the communities we serve are damaged by the failure to achieve sustainability objectives. There is no doubt about the threat to fiscal, environmental and human health if we maintain the status quo. Healthcare is part of the problem.

In every community, healthcare can be part of the solution. Healthcare's stake is unique. Although unable to independently battle the forces that contribute to the problem, healthcare can be a model of response at the local and societal levels, a source of expertise and a catalyst of action for a sustainable future.

Are the full implications of healthcare sustainability well known? Certainly not by all. Many healthcare organizations will be content with traditional roles of community benefit. Traditional community benefit is an immensely important contribution, but at a level of engagement that falls short of potential contribution and community responsibility. The achievement of sustainability requires a broader engagement by healthcare.

The role of the hospital is evolving as medicine advances both scientifically and organizationally from being the curative center to being the center of health. The new reality is that the hospital institution has a unique responsibility to address the challenges of sustainability to the extent that the hospital is responsible for community health. Hospitals must reach for a new definition of the health of the community, one that goes beyond the traditional measures of community benefit. The hospital is the only institution that has the mission, intellectual resources, reach into centers of influence, ability to command attention, and leadership potential to help guide the community toward sustainability. That is the full implication of sustainability for the hospital and the basis of a new social compact.

The mission of this book is to catalyze transformational change in the healthcare industry through sustainability. This book is not about tree hugging, nor is it about affection for Mother Nature, though there is a time and place for that. There is more to sustainability than environmentalism. The environmental movement does, however, provide a platform for the emergence of sustainability as the new mantra that integrates economic, political and moral forces to address the threats to the environmental dimensions of well-being. The essence of the new compact is clear. Through sustainability, healthcare renews its commitments to prevention and wellness, strengthening its compact between provider and community. Converting this compact into explicit, achievable objectives that address sustainability implications for a given community is a long, complicated process. It requires informed, thoughtful and committed professional healthcare administrators to focus institutional attention and energy on the consequences for the hospital and the health of the community.

The challenge of sustainability mandates an expansion of the scope of health administration professionalism

What value does the healthcare organization add to the community? It is in the answer to this question that the role of healthcare sustainability is addressed. If the value of the healthcare organization to its community is restricted to curative care and serving as a "doctor's workshop," then the hospital has what may be defined as a limited value framework (the hospital performs quality medical care). The administrator with a limited value framework says, "We are not in the business of healing the environment. We take care of sick people." The weakness of this limited value framework is that it does not align with the community's health needs. Making people whole again is an important part of healthcare needs, but it is only one dimension of a community's broader health needs.

How does the broader value framework get built into the DNA of the hospital? There is at least one set of people with the perspective to look across resources and into the community, to envision a broader value impact on the community, which we call a healthy community. Not every hospital or healthcare leader will be prepared for this challenge, but most hospital administrators are positioned to aggregate skills and resources as valuable contributors to achieving the community's healthcare sustainability goals. Executive leadership sets the tone.

If you visit an average town of 30,000 people and look around, you will find that a hospital is an indispensable community resource, woven into the fabric of community life. Every community depends on access to a hospital. The person who runs the hospital is probably ideally positioned (due to status, access to resources and place in the community) to improve the quality of life in terms of health and well-being, to create sustainable conditions in which people are most likely to flourish. Healthcare leadership has an opportunity to be the first citizen of sustainability because it is a dimension of healthcare. It can be a calling but above all it is a mandate: to be the person who raises the issue, facilitates the conversation and brings together the players. The hospital becomes a model, a demo for the sustainable citizen from which the entire community benefits. That role requires visionary leadership to bring a set of values and a sense of values not only to, but for, the community. Sustainability is central to the concept and definition of healthcare, and is therefore the business of healthcare and healthcare leadership.

The journey

This book is about the first steps in the journey of one hospital and its leadership team to lay the foundation of the new compact known as healthcare sustainability. The fabled Memorial Hospital is an important model of an organization in transition. The hospital integrates the full spectrum of services as the primary health resource for a large suburban and rural population. It has the respect of the citizenry, the support of community leadership, and well-established relationships with educational institutions, government and public health agencies. It is, by most measures, a successful healthcare institution, and its actions and activities are highly visible. The hospital is an adopter of environmentally friendly processes and procedures, but not without roadblocks and hiccups along the way.

This sustainability story follows the journey of Fred, Chief Executive Officer of Memorial Hospital, whose first executive role is as the newly minted hospital administrator. In the unrefined text of a narrated story, Fred provides vision and leadership for Memorial Hospital to leap beyond society's primary healthcare expectations to become a locally celebrated and nationally recognized healthcare sustainability leader. The organizational and leadership components of each chapter bring Fred one step closer to achieving the new provider–community compact, one that internalizes healthcare sustainability for community benefit. Achieving this mission requires informed and thoughtful commitment from Fred to focus institutional attention and energy on the consequences for the hospital and the health of the community. This book tells the story of the first steps in the journey of Fred and Memorial Hospital to describe how engaged, empowered teams advance healthcare sustainability objectives.

This book is a healthcare leadership resource for translating healthcare sustainability from governance to operations. It provides healthcare decision-makers with the information needed to navigate an ever-changing landscape, without being a dogmatic manual. The value-based principles exposed in this book are timeless.

This is a good thing, because the co-authors hoped a second edition of this book would not be necessary. While many healthcare organizations have progressed in the pursuit of healthcare sustainability since the first edition of this book was written, other organizations regrettably have not begun their healthcare sustainability journeys.

The *eureka* moment for any leader is that moment in which we identify the right question to ask. Like any great journey, this book will guide you to ask meaningful questions rather than take you to an answer. The book is about the thought process of sustainability. It is less about the details of a sustainability program and more about the big picture: sustainability as a vehicle for achieving leadership priorities.

Sustainability will open doors if you, the leader, ask the right questions. The first step on this journey is to have an open mind to embrace a future that is about health, not just healthcare. Sustainability is about the triple bottom line: not just cost, not just the community, not just the environment, but a combination of all three component parts. A healthy hospital strives to create a healing environment for patients and a healthy workplace for staff and the community it serves. Sustainability is about physician and employee engagement, cost-effectiveness and risk management, not just "going green." It is about wasting less, which inherently means saving more. Sustainability is about managing existing goals and transforming the way we think about the future of health.

The healthcare sector increasingly recognizes the critical link between human health and environmental health. Notable health systems such as Kaiser Permanente and Catholic Healthcare West, among others, paved brave inroads at a time when sustainability was not yet valued as the cover story. It is not just the large systems that are influencing sustainability in healthcare. Gundersen Lutheran, for instance, is one of the most forward-thinking sustainable healthcare organizations in the industry. Today, hospitals across the country and around the world, in diverse settings serving diverse populations, are developing organizational sustainability programs. Such programs build positive reputations by reducing the environmental impact of facilities and operations while concurrently saving money and minimizing liability and compliance risks.

Sustainability programs have the potential to reduce costs while protecting patient, staff and public health and well-being. Hospitals are the epicenter of the healthcare world as it exists today and are concentration points of care delivery. This concentration of care in one location also concentrates the environmental footprint of hospitals. Hospitals in the United States generate more than 5.9 million tons of waste each year,[1] are the second-largest commercial users of energy (behind commercial food service),[2] and use toxic chemicals in the process of treating patients, processing lab results and cleaning floors. Emerging science documents the growing effects of low-level chemical exposures on infants, children and fertility. A study in *The Lancet* reveals that nurses have the highest incidence of work-related asthma of any occupation studied, followed closely by cleaning staff.[3] This book demonstrates that all of these challenges and others can be addressed and reversed through sustainability efforts.

Sustainability is no longer an emerging trend: it has partnered handily with mainstream business frameworks of increased efficiency and process improvement to demonstrate ongoing value to companies and organizations that build sustainability into standard operating procedures. Siemens is one example of a company that champions sustainability as a core business differentiator. While sustainability programs are being integrated across the business spectrum, perhaps nowhere is the focus more appropriate than in our nation's challenged health system.

One dimension of this changing perspective means recognizing why and how sustainability permeates throughout every aspect of hospital operations. Leadership around healthcare sustainability began slowly. In the mid-2000s, however, as the drumbeat for sustainability grew stronger in the corporate world and the early pioneers in the healthcare space demonstrated that they could reduce environmental footprints while – surprise – saving money, the number of hospitals engaging in sustainability work began to climb. Today, healthcare facilities leading the way in environmental excellence range from rural, critical-access hospitals in Montana to sprawling metropolitan hospitals in the medical mecca of Boston. Facilities across the spectrum share an important common thread that is integral to their successes: senior-level buy-in and engagement, ongoing support and leadership. Healthcare executives play a critical role in helping healthcare sustainability programs grow to fruition.

The healthcare community is no stranger to change. Hospitals are constantly faced with the challenge of how to communicate effectively and sustain change within the multitude of systems involved in delivering excellent patient care. When a healthcare organization commits to the concept of environmental sustainability, it is embarking on change within its culture. To be effective, cultural change requires executive champions, and is most successful when the rationale for change resonates with staff at all levels. Creating a culture that embraces and integrates sustainability is driven by leadership, coordination, education and accountability.

This book provides indispensable resources to demonstrate why hospitals should consider prioritizing an early and aggressive sustainability strategy with clear, measurable and realistic goals and objectives. We developed this book to assist senior management in recognizing the value added by sustainable activities. Most of all, this book will provide the necessary tools to enable healthcare leaders to weave sustainability into the cultural fabric of your organization.

Characters

Character	Role
Amalie DuBois	Sustainability Director
Denise O'Hare	Chief Financial Officer
Dr. Grisha Vladimir	Physician – orthopod
Dr. Jenny Bea	Physician – pediatrician
Dr. Mordechai Goldberg	Primary Care Physician
Dr. Shmuel Cohen	Physician – hospitalist
Felipe Gonzalez	Vice President of Materials Management
Fiona O'Malley	Director of Marketing and Communication
Fred	Chief Executive Officer
Kadine Johnson	Chief Nursing Executive
Lorenzo Lopez	Director of Environmental Services
Mr. Singh	Director of Engineering
Ms. Crumpmuffet	Executive Assistant
Melvin Monroe	Vice President of Design and Construction
Dr. Arundati Rupee	Clinical Department Chair
Dr. Leah Gehry	Internist
Dr. John Kusek	Chair of Obstetrics and Gynaecology
Mrs. Penelope Lobatnic	Patient
Dr. Benji Renata	Women's Health Primary Care Physician
Lucient Uguayo	Chief Technology Officer
Dr. Humphrey Hedge	Chief Medical Officer
Georgia Trohv	Surgical Services Nurse
Franco Popellero	Surgical Services Unit Director
Gregor Verug	Director of Safety and Security
Lori Prince	Vice President of Human Resources
Sandy Smith	Director of Facilities
Julie Schmooly	Patient
Liam Ludwig	Director of Food Services

Notes

1 Anon. (2010). Green lantern: how much trash does a hospital produce? *Washington Post* October 18. www.washingtonpost.com/wp-yn/content/article/2010/10/18/AR20101 01804465.html.
2 Energy Information Administration (2008). Energy Star for healthcare. www.energystar. gov/index.cfm?c=healthcare.bus_healthcare.
3 Anon. (2007). Exposure to substances in the workplace and new-onset asthma: an international prospective population-based study (ECRHS-II). *The Lancet* 370: 336–41. www.creal.cat/pdf/dossier_lancet_def_eng.pdf.

ACKNOWLEDGMENTS

Knox Singleton

I want to acknowledge the power of continuous learning in the process of trying to define leadership strategies that will move healthcare leaders toward healthier and more sustainable organizations and communities. The sources of that learning can be pinpointed primarily in my co-authors, whose curiosity and efforts to identify and understand the world of leadership amidst the evolution of science, public policy and human events is apparently insatiable. Being part of this team of likeminded authors has brought fresh energy to my own curiosity and learning daily. To Seema and Carrie, I am eternally grateful for being the source of whatever personal growth I have enjoyed through the process of producing this book.

I also want to thank the many members of the Inova leadership team who demonstrate daily a fresh commitment and passion for creating a healthier community and by doing so provide stirring examples of engagement-driven success to others in our field. Professional learning is often defined in its most basic form by the "case" method; that is, "c"opying "a"nd "s"tealing "e"verything from those who have the courage and passion to be thought leaders and early adopters of effective sustainable leadership principles and practices. For your example, I appreciate the opportunity to learn and grow with you. Serving others with you has been a distinct honor.

To my family and friends who are like family, thank you for your ongoing love and support.

Seema Wadhwa

I am so blessed [...]

To have parents who have supported me, challenged me and most importantly loved me. I aspire to do more, be more and give more based on your example.

To have sisters who have blazed trails that I can follow. Your courage inspires me to take the trail less followed.

To have brothers who always encourage me and are there for me. Your passion for caring moves me.

To have nieces and nephews who always bring a smile to my face. You see the beauty the world has to offer.

To have mentors and friends who have taught me, believed in me and guided me. Your guidance continues to help me grow every day.

To have co-authors who turned into family. Your friendship warms my heart.

To have the most kind, amazing and supportive husband who has loved me. You have touched my heart and defined my future.

To have a beautiful baby girl through whom I get to see the world through fresh eyes. You drive me to make the world a better place.

I am so grateful to be so blessed.

Carrie Rich

Thank you to my children; the contentment of my life and your father's life depends on the hours we spend alongside you. We love you. You are our joy and hope, our happiness and the great loves of our lives.

Thank you to my superhero, home team.

Thank you to my parents, for making the burden of life's transitions as wonderful as possible. I am blessed.

Thank you to my colleagues, for making each day a gift.

Thank you to my village, for supporting me as I attempt to live my full potential and give back to the world.

1

HIPPOCRATES WAS RIGHT

FIGURE 1.1 What is sustainability?

A told story

FRED: I think, probably, the most painful moment of my life was when I had to
go in front of the Chairwoman of the Board at my first big-time job as the
newly minted Chief Executive Officer (CEO) of Memorial Hospital. I
followed a seasoned CEO who served 35 years, since the hospital was founded.
I mistakenly thought that a change in leadership and fresh perspective would
be a welcomed, easy undertaking. And there I found myself, confessing that
I didn't have all the answers to the concerns which the Board expressed.
Little did I know that the prior CEO had actually retired ten years before
his announcement was made. In fact, just about every major stakeholder of
the institution had lost faith in the organization. That is the backdrop in
which the Chairwoman of the Board requested that the hospital "go green" as
a top priority. As the CEO, "going green" was not exactly at the top of my
priority list. I just couldn't imagine how going green would impact any of the
dozen problems that I thought were in the way of the institution going
forward.

The Chairwoman didn't seem to understand that we were in a life-or-death situation as an organization. We were an institution that couldn't borrow money. We had lost the confidence of our admitting physicians, had the highest employee turnover rate of any hospital in the county, and we were generally thought of – by both elected officials and local business leaders – as somewhere around the World War I era in terms of our corporate responsibility. Ignoring all that, the Chairwoman wanted me to focus on "going green." Give me a break! I was doing all I could just to survive. Forget the sustainability stuff.

Where did sustainability come from anyway?

The current sustainability movement originated with the World Commission on Environment and Development (more commonly known as the Brundtland Commission), which was convened by the United Nations in 1983. In its report, *Our Common Future*, the Commission defined sustainable development as "development that meets the needs of the present without compromising the ability of future generations to meet their own needs" (1987, p. 43).[1] The Commission established that environmental crisis is typically broken down into a complex system of interconnected problems. These problems are caused primarily by human activities that overburden the Earth's natural environmental capacities. Localized problems concerning air quality, water quality, soil depletion, species loss, habitat loss, food production and energy supply are often related to one another in falling-dominoes fashion. Physical environmental problems are almost always linked in a vicious spiral with specific social conditions. Economic, political and cultural factors drive changes to the physical environment and affect the Earth's natural systems. Changes to those systems, in turn, force often dramatic and sudden changes in human society. For instance, crude oil and refined fuel spills from tanker ship accidents in the Gulf of Mexico, France, Kuwait Sundarbans and Ogoniland have devastating impacts on the ecosystem, population health and economic development, among other factors. Further, climate change, which was still largely a speculative hypothesis at the time of the Brundtland Commission's meetings, has emerged as one of the most prolific global dilemmas of our time. In short, environmental changes – both natural and manmade – impact community health and are reason for healthcare leadership to pay attention to sustainability.

(Bartels and Parker, 2011, p. 43)

What are the drivers of sustainability?

Understanding the drivers of sustainability is important for today's healthcare leaders, whether at the board, executive, middle-management, practitioner or trainee/educational level. Regrettably, there is often a paucity of understanding of sustainability and the variables that drive or influence its adoption in many, if not most, of these populations.

FRED: I'm hearing more and more of my peers say, "I think we need a sustainability program. We need to be involved." What's driving this sustainability movement? The truth is, I don't have the answers yet. So I phoned my friend and colleague Knox Singleton, Chief Executive Officer of Inova[2] in Northern Virginia. Inova hosted its first sustainability conference in 1992, long before sustainability in healthcare was commonly part of leader conversations. So Knox, why did you champion sustainability at Inova back in the day?

KNOX SINGLETON: As CEO of Inova, the drive to adopt sustainability as an organizational priority was largely a result of our employees; our employees then and now think about sustainability and really care about it. I've found that young people, especially, are increasingly educated or fired up about everything related to sustainability. Of course, the people who were young in 1992 are middle-aged today!

FRED: Knox, I notice the same thing. The younger employees behave as though sustainability is a given, or should be. And that may very well be. It's us "seasoned" and experienced folks, we're the ones who need to be educated about sustainability, it seems, not the other way around.

KNOX: I think that's a fair assessment, Fred. Generally speaking, young leaders believe the notion that environmental responsibility is a core value that they should practice in their personal and professional lives. To some degree, it's sort of a "secular faith" that all of us have a duty to address the environmental problems of the world. Just like service to others, sustainability appears to be a universally held value to younger generations.

FRED: I think I'm starting to get it, or at least be more open-minded to the idea of sustainability as a leadership priority.

KNOX: I'm sure you'd agree that part of the challenge in leading an organization is defining the higher order values that guide the purpose of why employees engage in the work we do.

FRED: Sure, of course. We've adopted the Studer model [Studer, 2004] of patient service for that very reason.

KNOX: Then your employees are familiar with Quint Studer talking about the two key attributes people want from their employment experiences. First, people want the ability to have an impact upon their work environment and team. Second, people want to feel like their work has purpose as part of a larger community.[3] They want their work to contribute to some purpose greater than their own self-interest. Some people find purpose in faith or a specific religion. Some people seek purpose by being a positive, good person, helping others and sustaining the world. Whether the affiliation is religious or not, these values transcend human differences; they are core, high-level values common to most everyone.

FRED: [*Gazing out the window, rapping his pen on the desk*] I sure have a lot to think about.

KNOX: Me too. I'm still learning about sustainability. The challenge remains how to meld health improvement with engagement around sustainability into a holistic leadership style.

FRED: Well that's good to hear, to be honest. This sustainability stuff is still a bit disconnected for me.

KNOX: I think that if you look at the core of healthcare, beyond making people better, it's rooted in the Hippocratic Oath to "First, do no harm." Today, many healthcare employees feel that purpose in responsibility for care of the planet. They believe in equity through shared use of natural resources. Remember: the principal reason why I embraced sustainability as a leader was that I felt it was one of the core values of many, many people who work within our healthcare family. Many employees understand that failure to use resources responsibly can have unintended, negative consequences on the environment and health of the community. Sustainability is in line with what Hippocrates had in mind.

THE STUDER MODEL

Two aspects of work that employees value about their employer:

- People want the ability to have an impact upon their work environment and team.
- People want to feel that their work has purpose as part of a larger community. They want their work to contribute to a purpose greater than themselves.[4]

WHAT IN THE WORLD IS A TABLE OF TABLES?

A table of tables is a decision-making tool used to prioritize the values and drivers of a given industry and/or market sector. This leadership tool may be used to evaluate the case for healthcare sustainability. The sustainability-leadership table of tables is one such tool that resulted in clear patterns, ultimately creating vision alignment for this book. Vision alignment achieved through the table of tables frames sustainability concerns in the context of a framework that resonates with healthcare leadership.

The sustainability-leadership table of tables

Sustainability may be used as a tool to frame leadership priorities, and ultimately may be used to enhance alignment among stakeholders around the promotion of health. Table 1.1 outlines key healthcare priorities from a leadership perspective as well as key sustainability drivers from the perspective of a healthcare sustainability director.[5] A holistic view of leadership requires an integration of these two sets of values/drivers. That alignment is the premise upon which this book is based. In short, *sustainability is a vehicle for driving leadership priorities.*

TABLE 1.1 Leadership priorities related to sustainability drivers

Chapter	Leadership priority	Sustainability driver	Operational topic
1	What is sustainability?		
2	Governance to operations		
3	Physician engagement	Physician engagement	Health information technology
4	Process improvement	Process improvement	Waste
5	Employee engagement	Employee engagement	Transportation
6	Patient satisfaction	Patient satisfaction	Food
7	Cost savings	Cost savings	Energy
8	Safety	Safety	Chemical management
9	Quality	Quality	Environmentally preferable purchasing
10	Growth/brand	Growth/brand	Green building
11	Community benefit	Community benefit	Water
12	Healthcare ethics – the future of health		

What should every healthcare leader know about sustainability?

FRED: I have no idea. I'm just figuring out what sustainability means. I'd better ask Knox. [*Dials phone*]

KNOX: [*Phone rings*] Hello. Knox Singleton.

FRED: Hello, Knox. This is Fred. Got a few minutes for a little advice? The Chairwoman of the Board keeps sending me her ideas relative to sustainability and I don't know where to start in terms of answering her questions. Frankly, the questions she asks lead me to more questions rather than answers.

KNOX: Okay, Fred. Well that's a good thing. That means you're thinking. What's one topic or question that's giving you trouble?

FRED: I can't figure out what I need to know about sustainability as a leader. What does sustainability mean to *you* as a healthcare leader?

KNOX: I would say that every healthcare leader should appreciate where sustainability falls in terms of customer expectations. At some level, the leader's core responsibility is to understand who her customers are and what their expectations are of her. Customers include patients, clinicians, physicians, community, employees and payers. And then there are also the elected officials. It boils down to understanding what all customers expect in terms of sustainability performance. Then the leader needs to connect those expectations into the strategic plan, budget and culture. That may be quite a lot to learn, actually.

FIGURE 1.2 How does climate change impact health?

MANAGEMENT TIP

In general, what every healthcare leader needs to know about her customers falls into two categories:

- What do customers expect?
- How do I deliver the desired benefits in ways that are effective?

The leader begins by identifying the benefit(s) she wants the customer to experience, such as compassionate and effective treatment for patients, a good place to work for staff, a willing partner and advocate for the community.

Keeping the main thing the main thing

FRED: I understand that even though a program or service may be valuable, that doesn't make it an organizational priority. So I'm still a bit confused as to why I should put sustainability high on the list.

KNOX: There are four overarching reasons why every healthcare leader should care about sustainability. First, the expectations of a lot of customers can't be ignored. Responding to customers is an essential part of what we do as leaders. Second, there are regulatory consequences with regard to sustainability. Several sustainability activities are regulatory requirements: LEED, mercury elimination, waste management practices, etc.[6,7,8,9]

FRED: Breaking the law, or showing up in the news over some public violation of dumping regulations, would not be a pretty picture for me or for Memorial Hospital!

KNOX: Right. In that case, the leader faces liability and legal exposure, in addition to the public relations exposure of failing to protect the health of the public and the environment.

FRED: Like the nuclear industry – that industry has a checkered image because it's sometimes perceived as not having a robust enough safety or environmental record.[10] For the healthcare leader especially – with accountability for public health and community resources – there's liability associated with not aggressively pursuing a sustainability program. That's clear to me.

KNOX: Third, – Am I boring you yet?

FRED: Maybe a little, but keep going!

KNOX: Okay, third, I expect for demographic reasons we'll face another healthcare workforce shortage in the years ahead.[11] For most new healthcare workers, a significant consideration in their employment decisions includes working for organizations that they feel embodies their principles, their belief systems.[12,13] Healthcare workers will have choices between multiple places to work, and sustainability credibility is a factor when choosing a new employer. Fourth, quality management and cost-effectiveness occur primarily in organizations with cultures that embrace process improvement; employees have to be committed to driving waste out of work processes. People who think Lean and act Lean, for example, will naturally be drawn to the sustainability mindset around resource utilization and waste. When deployed effectively, both Lean and sustainability lower long-term operating costs and waste less in an economic sense.[14,15]

THE ROLE OF THE BOARD

The Board of Trustees plays a principally fiduciary role. It protects the assets that have been entrusted to the organization. To some degree, the Board also protects the community's interests so that the assets are utilized appropriately.[16] Governance is largely about disposition or strategic direction, which usually takes the form of supporting or redirecting the leadership advice of management. Boards usually do not generate strategy, nor do they invent the mission of the organization. They certainly do not create new tactics; at least, good boards do not. The Board's principal job is to keep management aligned with the organization's basic purpose and mission while they mitigate risk in terms of protecting the organization's assets. The two key responsibilities the Board has are to:

- Protect the assets of the organization.
- Protect the mission of the organization.

FIGURE 1.3 From the boardroom

Translating sustainability values from the boardroom to the management team and into the community

A healthcare leader's responsibility is a 360-degree responsibility. One fundamental responsibility of leadership is to articulate constituents' and customers' expectations to the Board. For instance, patients and the community generally have a clear expectation of a safe environment within a hospital. The patient expects not to be harmed by infections or by incompetent staff while at the hospital. Employees, visitors and patients expect the physical environment of the institution to be safe.

Most healthcare organizations embrace the principle of improving the health and quality of life of individuals and communities. This principle is consistent with embracing sustainability. There is one respect in which sustainability conflicts with the traditional healthcare culture, however. In the treatment-oriented model that dominates healthcare today, more is better. Resource-consuming processes, such as staying in the hospital longer and frequent use of diagnostic procedures, are often equated with perceived better care or outcomes among patients. In sustainability, less is more.

Creating a culture of sustainable health: top-down, bottom-up

The new paradigm of health will be driven by a culture that gets paid for healthy results rather than for producing more treatment.[17] While sustainability may not be perfectly aligned with today's traditional, treatment-based healthcare model, it is a valuable tool for shifting the healthcare culture from its historical roots to a more healthful, society and environmentally friendly cultural value set going forward.

The Chief Financial Officer needs to understand and address how to integrate sustainability into the financial program. The Chief Nursing Executive needs to understand how sustainability integrates with nursing practice. The support service professionals must understand how sustainability drives equipment and supply selection, food purchases and other logistics.

At every level of the organization, the healthcare leader holds colleagues accountable for driving the integration of goals deep into the organization and helps every employee to translate sustainability values into their work. Sustainability goals should be driven like financial goals, safety goals, patient experience goals – these goals will be driven down through department levels and pushed up through grassroots engagement. Fully embraced sustainability goals are driven down to the staff nurse and the billing clerk, and up to the executive leader. Sustainability requires a top-down *and* bottom-up culture-change approach so that leaders at all levels of the organization are engaged.

In a healthcare organization that embraces sustainability, employees at every level of the hierarchy consider themselves environmental agents, and the healthcare organization considers itself an educational resource for the community to learn about sustainable practices.

EDUCATING LEADERS

The importance of educating current healthcare leaders about sustainability should not be underestimated. For many rising healthcare leaders sustainability is part of their core ethos, part of their upbringing. But for many "experienced" healthcare leaders sustainability has not been part of their training, and in some cases may not be part of their personal experience or value system. And yet, sustainability is central to the role of healthcare leaders of all generations – to be informed about sustainability, to understand it, to be able to represent it, and to achieve positions of organizational clout while embracing sustainability as an ethical responsibility and a moral imperative.

Why is sustainability particularly important for healthcare as compared to other industries?

Sustainability is particularly important for healthcare for two reasons. First and foremost, healthcare expectations are to "first, do no harm." Failure to meet the mandates of sustainability results in harm to the community. Because sustainability is an integral dimension of health status, it is implicitly or explicitly incorporated into the mission of the organization. Second, healthcare is a huge resource consumer.

KNOX: The provision of health services is a resource-intensive, inherently hazardous activity. In short, healthcare is a resource hog and prolific generator of solid waste. Healthcare is a large and growing segment of employment in the United States and our overall economy. We consume huge amounts of energy, water and other resources. The care process produces a large amount of potentially polluting and dangerous by-products. We share the environmental impacts of many other large employers, such as employee transportation. We understand our impact in terms of the present and future health of our community. That puts a special responsibility on us.

FRED: Of course. That makes sense.

KNOX: So, above all, both sustainability and the emerging wellness paradigm focus on culture, mission and values centered on doing only those actions that benefit patients. Sustainability embraces the concept of examining consequences and preventing problems, avoiding downstream sequelae – like unnecessary waste, energy misuse, air pollution and water contamination. Hospitals and healthcare organizations have enormous impacts on communities and the environment. They have large numbers of employees who drive to and from work every day, just to give one example.

FRED: And I guess, like all drivers, they text while driving in construction zones, and have accidents, and place additional burdens on the system.

KNOX: [*chuckling*] Well, that's one way of looking at it. Another perspective is that healthcare is a huge producer of pollution into our natural systems. We detrimentally impact our environment – water, traffic, education, commerce – all

FIGURE 1.4 Healthcare and climate change

of which in turn may have a negative impact on public health. Sustainability, at its core, promotes that preventive, do–no–harm mindset. How do we as a health system perform in those areas of public health impact? Many hospitals today advocate against second-hand smoke and for a variety of policies and practices that improve the environment for patients. That's one way in which healthcare delivery links to public health. Sustainability in healthcare tends to focus efforts on getting the fundamentals right. So, if in healthcare delivery we refocus upstream on prevention rather than treatment, life-cycle rather than upfront costs, then by default we embrace public health concerns. Let's face it, on every healthcare leader's priority list is how to engage physicians around the important priorities of the institution. Leaders are trying to figure out how to integrate physicians into healthcare teams, especially younger physicians, who are often more familiar with new processes and approaches. Connecting with physicians is important for redesigning clinical processes to drive out wasted steps and reduce costs. So we're seeing that the processes of waste reduction and physician engagement line up. They both tie to and parallel sustainability.

FRED: That's an eye-opening way to think about physician engagement. And, of course, you're not only talking about physicians. It applies to all of our employees. It's a new culture.

ABOUT THIS BOOK

This book concentrates on how to focus the attention of the leadership team on sustainability. Frankly, nothing much on the sustainability agenda tends to be on the average healthcare leader's priority list. Process improvement, meanwhile, is on almost everyone's list. The notion of driving waste out of the system is a starting point where healthcare leader priorities link to sustainability. There is a lot of interest in how to reliably drive waste out of healthcare processes while continuing to deliver necessary, high-quality medical services; how to be cleaner, leaner and now greener. These are key concepts on which the sustainability enterprise is focused.

KNOX: Fred, I wrote a book about the intersection of sustainability and healthcare leadership, a book about how integrating sustainability within our organizational culture has enhanced my team's effectiveness. It makes a great holiday gift!

FRED: Knox, it sounds like you're drinking the Kool-Aid.

KNOX: [*laughing*] The crazy thing is, though, that I wasn't so into sustainability originally. I mean, I understood its value in terms of employee engagement and the other leadership priorities we've been talking about. But I didn't realize how important it would be for me to build sustainability within the scope of our leadership approach and style.

FRED: It sounds like you were in my shoes when you started your sustainability journey.

KNOX: That's exactly right, Fred. When we began writing this book, I perceived the case around cost structure to be a little weaker than the other sustainability drivers. "I'm going to have to disagree with you about cost impact," our Sustainability Director told me. "We've saved over $100,000 in the past two months alone through reprocessing medical devices." I was shocked that we had saved over $1 million that year through another sustainability program. I wasn't up to speed on the financial impact of our sustainability efforts. But Fred, I want to leave you with a different perspective that is starting to emerge for me.

Our engagement with sustainability has taught me several lessons, and I'm sure there will be more as we proceed on the journey that is the basis of this book. One lesson is that sustainability is an essential component of the emerging wellness paradigm that results. Another lesson is that sustainability is a unifying element of our organizational culture. Everyone appreciates the importance of "do no harm" as employees try to do good for patients and the community. The ultimate engagement reward is for everyone in the organization to act as stakeholders in sustainability.

The key to this book is the concept that if you, as a healthcare leader, care about physician recruitment and engagement, process redesign, employee engagement, facility investments and cost structure, then let us tell you about why you should care about sustainability. Each chapter addresses top priorities that healthcare leaders care about. Exploring each leader priority and its link to sustainability impresses our main character, Fred, the healthcare CEO. Every step we take toward linking healthcare values with sustainability drivers is intended to enrich leadership around the value of sustainability as we move toward a new paradigm of health.

SUSTAINABILITY IN ACTION

Sister Susan Vickers is the Vice President for Corporate Responsibility for Dignity Health, a health system composed of more than 60,000 caregivers and staff who deliver care in 21 states. Headquartered in San Francisco, Dignity Health is the fifth largest health system in the nation and the largest hospital provider in California. She is responsible for directing and overseeing system-wide sustainability and corporate social responsibility programs. Susan currently serves as a board member for Health Care Without Harm, Practice Greenhealth, Mercy Investment Services and the St. Mary Medical Center Foundation. She has also worked as pastoral minister in assisted rental housing for elderly and disabled individuals, and spent 20 years teaching and administering in elementary schools.

What spurred your interest in environmental sustainability?

It goes back 20 years, to 1996. There was a confluence of factors that caused Dignity Health to move the needle on sustainability. A major factor was that nurses were concerned about the waste we were continually generating.

A second factor sprang from our roots as a religiously sponsored healthcare system. Congregations wanted to address environmental concerns through their ministries in each of our eight sponsoring communities. At the same time, the executive management team at Dignity was pushing in the direction of sustainability.

We worked with Ceres – a nonprofit organization working with investors and companies to build leadership and drive sustainability solutions throughout the economy – to draft a set of environmental principles to adopt as policies. We then created an annual report to guide and monitor our progress toward achieving sustainability goals. And then we met Gary Cohen who had just founded Health Care Without Harm (HCWH). We were the first healthcare system to become members of HCWH and to endorse its main goal of reducing the volume and toxicity of waste in healthcare.

What key levers have you used to ingrain commitment to sustainability into Dignity Health's culture and business practices?

The link between our mission and health promotion is a key lever. We try to tie everything we do to this mission. The policy commitment on the part of the Board, to set sustainability goals on an annual basis and report progress to the public every year, holds us accountable.

In addition, identifying and supporting champions at various levels of the organization has been a key lever. Environmental sustainability is a priority for many of our employees, and many employees are eager to advance sustainability best practices in the workplace.

How do you build sustainability into your leadership?

We had a major breakthrough over the past two years in terms of executive leadership buy-in to sustainability. The executive team had always supported sustainability efforts, but rarely engaged directly. A couple of events sparked greater engagement and ownership. First, Dignity Health was invited to speak with other HCWH members about the nexus between climate and health at meetings held concurrently with the Paris Climate Talks. This participation captured the imagination of others at Dignity who hadn't been paying much attention before. Second, in preparation for our 2016 sustainability report, we conducted a comprehensive ESG (environmental, social, governance) materiality assessment. As a result of the assessment we established a system-wide governance structure for sustainability. Now we have a cross-functional, system-level council that is focused on integrating sustainability throughout the entire organization.

What tools or resources have you found helpful in your journey toward sustainability in healthcare?

HCWH and its subsidiary Practice Greenhealth (PGH) have proven to be valued partners. PGH is a membership organization with a mission to mentor and assist hospitals in assessing and addressing their environmental footprint. In the past few years they sponsored a Healthier Hospitals Initiative (HHI) to develop challenges and toolkits for hospitals to implement to address six key challenges. This initiative has been a huge help for hospitals.

Most recently, a new project, Greenhealth Exchange (a buying co-op), has brought hospitals together to promote green healthcare products and aggregate the purchasing of these products, making them more affordable. This is a really promising development in the marketplace, as it encourages key players in the industry to purchase these products.

How has the healthcare industry changed as it relates to environmental sustainability?

In healthcare, we have been slow to understand our footprint and take action to address it. In the past few years there has been a rapidly growing under-standing of our impact and responsibility, development of sustainability tools for the healthcare setting, as well as the promotion of individuals who have a passion for sustainability. We are making solid progress at this point. However, we still have a long way to go. For example, only 1,200 hospitals out of 4,000 in the US are members of Practice Greenhealth. Among the 1,200 that are members, the level of commitment and achievement varies greatly.

Any advice for latecomers to sustainability?

Focus on your mission. Link sustainability goals to your mission. Don't start from scratch. Use the tools available to you at PGH, HCWH, and other organi-zations embracing sustainable operating practices. Start with achievable goals and build from there. Surround yourself with likeminded champions who col-lectively move forward sustainability in healthcare.

Practical application: leadership's role in implementing a sustainability program (Gundersen Lutheran)

The imperative

- Create a more sustainable environment for patients.
- Decrease the community's environmental burden.
- Use sustainability to promote organizational goals such as improved access to care and lower clinical error rates.

- Weave sustainability into the organizational fabric by linking sustainability to the organization's mission.
- Fulfill healthcare's moral obligation: "first, do no harm."
- Distinguish the organization as a leader in the field of sustainability.

The initiative

Gundersen Lutheran's spark for sustainability came from the CEO. In his previous role as Executive Vice President (EVP) he had dabbled in a few sustainable initiatives but found he had little time to pursue these objectives, given other demands on his time.

Once in the CEO position, he was able to organize an initial strategy team of volunteers comprising himself, a VP and a few colleagues who expressed interest in pursuing more sustainable operations. To create a vision of what it would look like to be a more sustainable organization, the sustainability team reached out to key internal leaders who were already involved in such activities. Shortly thereafter, they hired a dedicated resource whose charge was to improve the organization's sustainability approach.

The new hire reported to the Senior VP (SVP) of Business Services after joining the organization from an outside industry. The SVP brought diversity of perspective to the team and was open to new ideas, all while sharing common priorities of improved quality and lower costs. The CEO met routinely with the newly created sustainability position, the SVP of Business Services and the Director of Facilities to help organize their ideas and set the course for the organization's sustainability program.

The team created several core sustainability work groups that focused on topics such as energy, waste management and purchasing. The sustainability program slowly gained traction within the organization as the internal team created detailed plans and showed continuous progress. By starting small and demonstrating an array of successes, the sustainability team achieved buy-in from the Board to pursue larger projects, which eventually snowballed into the award-winning sustainability program that exists today.

The metrics

There are a variety of mechanisms to measure the qualitative impact, including employee engagement scores, patient satisfaction and financial cost savings, which will be discussed in detail in future chapters. In addition, qualitative feedback from senior leadership, board members, medical staff leads, patients, employees and senior community members are all excellent gauges of project success. Project integration is another key indicator of the speed of adoption within the organization.

Lessons learned

- There will never be a way to get *everyone* fully engaged – there will always be naysayers for one reason or another (often due to competing priorities).

So long as they do not try to obstruct integration efforts it is best to let them be, and focus engagement efforts on the individuals who want to be part of the sustainability program.

- Publicity helps garner engagement. When a leader highlights the value of a sustainability project in terms of what it means for both community and staff, it helps the leader and organization gain momentum to expedite other sustainability projects. Public and staff support is especially important to board members, the community and investors.
- Framing the sustainability discussion in the language of the organization's mission and responsibilities will help achieve buy-in. Make sustainability part of the work of the organization as often and as much as possible.
- Do not lead with a "we should do this" command. Explaining how other organizations have approached sustainability practices and the outcomes achieved will decrease knee-jerk reactions to being told what to do.
- Broaden employee and leadership education about what sustainability means by using a multi-pronged approach focused on outcomes to increase the likelihood of hitting a touch point.
- Find an executive in the organization to champion the sustainability cause: the more executive the champion, the better. Once this person is identified, use him or her sparingly to help promote sustainability at the appropriate times. He or she will be an important source of insights into the organizational triggers for accomplishing sustainability goals.
- Avoid the "bleeding edge" phenomenon. When starting out, it is important to build momentum by implementing sustainability practices that will really work. Demonstrating early successes will increase the visibility and credibility of the sustainability program, and increase the likelihood that leadership throughout the organization will be engaged in and excited about sustainability.
- Be clear from the beginning about what is expected of key individuals and the timeline for achieving goals.

About the organization

Gundersen Lutheran Health System is a physician-led, not-for-profit healthcare system located throughout western Wisconsin, northeastern Iowa and southeastern Minnesota. It is a comprehensive healthcare network including one of the nation's largest multi-specialty group medical practices, regional community clinics, hospital, home care, behavioral health services, vision centers, pharmacies, and air and ground ambulances.

Sources

www.gundluth.org.
Jeff Thompson. Personal interview, 2013.

Leadership imperative

The outcomes of improving sustainability go far beyond reducing costs, important though cost is. Sustainability outcomes are a form of community service boasting a long list of positive impacts for society, the community and individuals. Sustainability is a core value of our individual and collective professionalism in healthcare. It is appropriate for our institutions and our communities to look to us for leadership in this domain.

Leadership dialogue

1   How can the healthcare leader who wants to learn more about sustainability get up to speed?
2   What are sources of information on the technical sustainability topics that will be most useful for a healthcare leader?
3   Establishing priorities is a key responsibility of leadership. Which sustainability priorities align with the priorities of healthcare leaders?
4   What, if any, are the benefits and risks of endeavoring to educate the board about the "why" and the "how" of engaging an organization in sustainability efforts?
5   Which values of health service delivery organizations align, or conflict, with the values of sustainability?

Syllabus integration

Overview

The United States healthcare system requires multidisciplinary approaches to address the dynamic field of healthcare sustainability. This lesson explains why healthcare sustainability is important. Students learn about frameworks for sustainability, identify stakeholders and gather evidence-based resources for sustainability management.

Objectives

- Introductions.
- Review course syllabus.
- Introduce patient, worker and environmental safety in sustainability terms.
- Identify key stakeholders and information repositories.

Topics

- Issue framing.
- Societal interest in sustainability and healthcare.
- Current state of knowledge.
- Healthcare's environmental impact.

- Corporate responsibility and social entrepreneurship.
- Triple bottom line.
- Community benefit.
- Legal considerations.

Homework

Advisory Board, the (2015). *The Health Care Sustainability Initiative.* Available at www.advisory.com/community-impact/health-care-sustainability-initiative.

Global Health and Safety Initiative (2009) .*The Eco-Health Footprint Guide: Measuring Your Organization's Impact on Public Health and the Environment*, version 1.2. Available at www.globalhealthandsafety.org.

Practice Greenhealth (n.d.). *Special Report: Building the Business Case for More Sustainable Hospitals.* Available at https://practicegreenhealth.org/about/press/blog/special-report-building-business-case-more-sustainable-hospitals.

Review the mission statements of healthcare organizations committed to sustainability. Create a mission statement for your imagined or real healthcare organization. Record your mission statement in your career journal.

Additional resources

American Hospital Association – Signature Leadership Series. *Environmental Sustainability in Hospitals: The Value of Efficiency.* Available at www.hpoe.org/Reports-HPOE/ashe-sustainability-report-FINAL.pdf.

Kaplan, S., Orris, P. and Machi, R. (2009). *A Research Agenda for Advancing Patient, Worker and Environmental Health and Safety in the Healthcare Sector.* Global Health and Safety Initiative, October. Available at www.globalhealthandsafety.org.

www.resilience.org/stories/2012-08-03/sustainable-healthcare/.

Notes

1 Brundtland Commission (1987). *Report of the World Commission on Environment and Development: Our Common Future*, p. 43. Available at www.un-documents.net/wced-ocf.htm.

2 Inova is a not-for-profit health system based in northern Virginia. Governed by a voluntary board of community members, Inova's vision is to optimize the health and well-being of each individual. Inova is reinventing itself around predicting and preventing disease, individualizing therapy and wellness, and transforming patient care through genomic research.

3 Studer, Q. (2003). *Hardwiring Excellence: Purpose, Worthwhile Work, Making a Difference.* Gulf Breeze, FL: Fire Star Publishing.

4 Ibid.

5 Practice Greenhealth (2012). *Sample Job Description: Healthcare Sustainability Director.* Reston, VA: Practice Greenhealth. Available at http://practicegreenhealth.org/pubs/toolkit/greenteam/SampleJDSustainabilityDirector.pdf. Rich, C., Singleton, J. and Wadhwa, S. (2011). Teaching sustainability as part of healthcare management studies: challenges, best practices and case studies. CleanMed, 2011, Phoenix, Arizona, April 5–8, 2011.

6 Department of Transportation (2011). Electronic Code of Federal Regulations, July 7. Pipeline and Hazardous Materials Safety Administration. Available at www.transportation. gov/pipelines-hazmat.

7 US EPA (2004). *Code of Federal Regulations Title 40 Protection of Environment, Section 311,* revised July 1. Washington, DC: Environmental Protection Agency. Available at http:// edocket.access.gpo.gov/cfr_2004/julqtr/40cfr311.1.htm.

8 Anon. (2008). Fairfax County Board of Supervisors adopts green building policy. Fairfax County Board of Supervisors, February 11. Available at www.fairfaxcounty.gov/ news/2008/030.htm.

9 OSHA (2011). *Hazardous Waste.* Washington, DC: Occupational Health & Safety Administration, US Department of Labor. Available at www.osha.gov/SLTC/ hazardouswaste/index.html.

10 Mufson, S. and Yang, J.L. (2011). A quarter of US nuclear plants not reporting equipment defects, report finds. *Washington Post,* March 24. Available at www.washington-post.com/business/economy/a-quarter-of-us-nuclear-plants-not-reporting-equipment-defects-report-finds/2011/03/24/ABHYa2RB_story.html.

11 Derksen, D. and Whelan, E-M. (2009). *Closing the Health Care Workforce Gap: Reforming Federal Health Care Workforce Policies to Meet the Needs of the 21st Century.* Washington, DC: Center for American Progress. Available at www.americanprogress.org/issues/ healthcare/reports/2010/01/15/7135/closing-the-health-care-workforce-gap/.

12 Block, R. (n.d.). Boomers: you need to rethink seeking full time jobs with Gen Xers. *Workforce50.com.* Available at www.workforce50.com/content/general_resources_ workforce50.html.

13 Lewis, A. (2011). How my company hires for culture first, skills second. *Harvard Business Review,* January 26. Available at http://blogs.hbr.org/cs/2011/01/how_my_company_ hires_for_cultu.html.

14 The economic benefits of sustainable design. Available at www1.eere.energy.gov/femp/ pdfs/buscase_section2.pdf.

15 http://planning.ucsc.edu/.

16 Buchbinder, S.H. and Shanks, N.C. (2012) *Introduction to Health Care Management.* Burlington, MA: Jones and Bartlett Learning.

17 Anon. (2011). At long last […] pay for outcomes starts to replace pay for performance. *Managed Care Magazine,* September. Available at www.managedcaremag.com/archives/ 1109/1109.payforoutcomes.html.

2

FROM GOVERNANCE TO OPERATIONS

FIGURE 2.1 Up to bat for sustainability

Executive summary

Sustainability is often discussed in terms of functional disciplines (energy, waste, water, etc.); this book discusses sustainability in the context of healthcare leadership. Leadership defined through sustainability provides context for strategy, and can be an accelerator of strategy. Leadership exercised through sustainability can be a framework for guiding cultural reinvention within an organization. Healthcare leaders who integrate sustainability into their priority objectives achieve successful outcomes across a portfolio of programs and services. Successful outcomes achieved through sustainability include improved community relationships, lower long-term operating costs and engaged employees. Sustainability is a mindset which many employees value and will rally behind, especially younger employees.

Governance dictates the strategic priorities of an organization. A core challenge of organizational leadership is the implementation of strategic objectives set by the Board, shareholders or government. External realities, such as sustainability, are a second set of imperatives that leaders must pursue. The leader's starting point for pursuing the Board's priorities and sustainability objectives is engaging executive colleagues to answer the questions "Why should I care about sustainability?" and "How can sustainability help achieve business success?" This chapter guides readers from understanding *why* leaders should embrace sustainability to understanding *what* the leader's role in healthcare sustainability should be. By the end of the chapter, readers will be able to make a compelling case for why their organization should invest in a sustainability program. Readers will also gain an appreciation for translating leadership priorities into sustainability tactics with beneficial operational outcomes. The organizing concept is that engaging the principles of sustainability can help healthcare organizations reduce long-term costs, improve outcomes and nurture the environment that influences health.

A told story

After the board meeting, Fred, CEO of Memorial Hospital, returns to his office perplexed. He's past the shock of the Board Chairwoman's suggestion of a hospital sustainability program. Fred wonders how to translate the theoretical sustainability construct from governance to operations. Fred decides the only way truly to engage the organization in sustainability practices is to engage the executive leadership team by hiring a Director of Sustainability to champion the cause throughout the hospital. Fred initiates a conversation with his leadership team on Monday morning. After several leadership team members identify other priorities that trump investing their time in a sustainability program, Denise, the Chief Financial Officer, agrees to be the executive sustainability champion.

FRED: Good morning, team. How is everyone this morning?

LEADERSHIP TEAM: Good morning [*voices lack energy, one person yawns while other colleagues chug coffee*].

FRED: I hope everyone had a great weekend. I'm delighted to be starting the week with you on this fine Monday morning. Let's go around the table and share our perceptions of the board meeting. I'm curious to hear your perspectives. [*Each member of the team shares his or her perspectives about the board meeting.*]

DENISE: [*Chief Financial Officer*] So, Fred, what's your perception of the board meeting?

FRED: You'll recall that we discussed the hospital's financial standing, our vision for the future and our strategy for achieving our vision – all the standard elements of governance conversation. But what really surprised me was when the Chairwoman approached me to discuss sustainability. Our Chairwoman was insistent about developing a sustainability program at Memorial Hospital. The more I think about the Chairwoman's rationale, the more I've come to realize that a sustainability program might not be such a bad thing. In fact, it might actually be a good thing, though I have no facts to confuse me. So, who's interested in leading the development of our sustainability program? [*Fred looks to Dr. Goldberg, Chief Medical Officer.*]

DR. HEDGE: This isn't a good time for me to take on another assignment. We've got Joint Commission coming soon and that's our focus. Besides, getting our medical staff involved in something like sustainability – which they know little, if anything, about – will be very difficult.

FRED: Okay then. How about you, Kadine? As the Chief Nursing Executive, what do you think about championing sustainability?

KADINE: I think developing a sustainability program is a wonderful idea. I really think you ought to lead it, Fred. You're fairly new here, and this is an excellent opportunity for you to be seen as a leader. [*Leadership team members roll their eyes at Kadine.*]

FRED: [*sighs and turns to Denise, CFO*]

DENISE: I don't see how sustainability fits with our priorities and financial strategy. I will say that my kids have been talking about recycling, which I know is related to sustainability. I believe that younger generations care about sustainability. We should think about how we can be proactive in the sustainability arena as we think about recruiting young talent.

FRED: Okay, thanks to all for sharing your perspectives. I'll go back to the drawing board and see what I come up with in terms of accountabilities. [*Fred reviews his notes after the meeting and decides that the majority of his team will follow his lead into a sustainability-based initiative, though perhaps with some trepidation. As usual with the engagement process, there is no "big bang" or dramatic moment of attitude change, but rather the beginning of an evolutionary shift. Each leader approaches sustainability from different perspectives with their own set of operating goals in mind. Leadership's understanding of sustainability and engagement around it is largely determined by those different perspectives.*]

FRED: [*thinking to himself*] I want to respond positively to the Chairwoman's direction to engage in a sustainability agenda. There's some wisdom in her argument that Memorial Hospital should be at the front of this parade. Our leadership values have to connect with operations. So, then, who are my choices to lead the effort? Dr. Hedge, Chief Medical Officer, is only thinking about leading the physicians through the Joint Commission review. He's distracted and unenthusiastic – not the way to introduce a new framework for leadership. I think the Chief Nursing Executive, Kadine, is just looking for a way to avoid the job. I guess I could think about designating her as our champion, but she didn't show much interest. And like a lot of managers, she'll get on any bandwagon, whether she cares about it or not. But I don't want a leader with so little enthusiasm; I want a *leader* who will make things happen.

That leaves me with Denise, my unlikeliest choice. She's the only one who's got genuine passion for sustainability. She sees the potential connection of sustainability to the community. I have to start somewhere, so maybe Denise is our executive champion for sustainability. I need to schedule a follow-up meeting with Denise. [*Before Fred schedules the meeting with Denise, he checks his email and finds the following note.*]

> Dear Fred,
>
> I know I'm "the numbers person," but I have been thinking about our sustainability conversation. I'm not sure yet how this will all play out, but I imagine we've got lots of employees who will likely be interested in a sustainability program. And then there's City Council – we're going to have to take our bonds to City Council next month, and they're always talking about ways to make our environment more sustainable. It's clear that sustainability is a big issue for the Chairwoman. I'd be willing to think through with you what sustainability might mean for the hospital and ultimately oversee the translation of sustainability from governance to operations.
>
> Perhaps you might consider connecting with Felipe, Vice President of Materials Management, to let him know that much of the ultimate operational responsibility will fall to him. I think this will be a leadership development opportunity for him. I look forward to discussing this important initiative with you.
>
> Sincerely,
> Denise
> Chief Financial Officer, Memorial Hospital

FRED: [*thinking to himself*] I'm glad Denise is stepping up to the plate. One of our key challenges will be answering the question, how do we engage employees? Our early adopters are the people with a passion and natural interest in sustainability. How do we inject that passion into the larger group? We find the natural leader who can be a role model that other people will relate to; we let her connect everywhere there is sustainability interest. We need early wins. Early wins will draw in less interested individuals at the outset. Maybe we pick two or three leadership principles we're trying to affect through sustainability, and introduce sustainability to stakeholders in the context of the selected leadership principles. I'll leave it to Denise to pick leaders who successfully engage employees on the broader Memorial team and in the community.

It's unexpected to have the CFO be the sustainability champion, but I think it's a blessing. After all, Denise will understand the fiscal implications of sustainability, and she oversees Materials Management, which will be responsible for the implementation of our sustainability program. Thankfully, Denise volunteered.

I think this whole sustainability thing has a future at Memorial Hospital.

The story of sustainability in healthcare: vision and alignment

A number of leading observers of the United States healthcare system have dubbed it "unsustainable."[1] Health outcomes are hardly outstanding when compared with other countries, especially given the high costs.[2] The level of resources consumed by US healthcare has grown steadily, rising from 9 percent GDP in 1980[3] to 17.8 percent GDP in 2015.[4] It is a system characterized by poor performance in the area of environmental sustainability as well.[5] From economic, social and environmental perspectives, the US healthcare system is full of waste, and replete with negative side effects for people delivering the care as well as the people receiving it. This scenario calls for healthcare leaders to pursue the "triple bottom line" of economic, social and environmental responsibility.[6] This call for action has permeated across the world to create a global network of hospitals all focused on reducing their environmental impact and creating a dialogue related to the relationship between environment and health: Global Green and Healthy Hospitals (GGHH). GGHH's agenda provides a comprehensive framework with metrics to engage the entire healthcare sector (HCWH, 2013).

WHO'S WHO IN HEALTHCARE SUSTAINABILITY

Sustainability in action

Mary Larsen is Director of Sustainability and Supplier Diversity at Advocate Health Care. She is responsible for directing, planning and managing Advocate's environmental stewardship initiatives. In alignment with Advocate Health Care's mission, values and philosophy, she sets the strategic direction and vision for responsibly mitigating Advocate's environmental footprint.

Advocate Health Care feature

Why is sustainability important to you?

I am motivated by mitigating the unintended environmental impact the healthcare sector has as a result of delivering healthcare. It is ironic that, in the process of caring for the health of our patients, we are inadvertently causing harm to the health of the environment. Healing patients on a sick planet is not sustainable in the long run. It is important to mindfully manage the resources we use for health and healing – the materials we touch, the toxic chemicals we use and the waste we generate – if we truly seek to improve the life expectancy and overall well-being for generations to come. To be true to our mission and ministries of health in the most holistic way possible, it is imperative that we conduct business operations with the future in mind as we solve day-to-day management problems.

What opportunities exist day to day to link environmental and personal health in your organization? How do you bring these opportunities to life?

Every day, we think of incentives and positive peer pressure opportunities to motivate our employees. For example, we encourage them to drink a lot of water from reusable bottles to minimize the waste of disposable bottles. Last year we distributed reusable water bottles to over 12,000 staff during an employee appreciation event. At our corporate office, we eliminated paper cups, napkins and utensils, and explained why they are bad for the environment, citing the environmental and economic savings. We provide fresh filtered water and coffee to employees and encourage them to bring to work their own reusable mugs and cups. We created a Sustainability and Wellness website to provide background on key initiatives, promote awareness, and provide an annual Healthy Environment Dashboard of results.

How are you working to link sustainable healthcare through the wellness initiatives at Advocate Health Care?

My team partnered with our internal wellness program, Health*e* You. Like many organizations, we invest in creating a healthier workforce by raising awareness about healthier personal habits. About three years ago, I partnered with Health*e* You to raise awareness about the inextricable links between personal and environmental health. We created a recognition program and ongoing incentives for employees who participate in the health plan. By participating in educational events or various healthy behavior promotions, staff members increase their chances of winning cash prizes every quarter. We provide examples of educational topics that individuals can pursue to achieve dual health and sustainability benefits, such as:

1   Using stairs.
2   Eating less meat and meat not raised by the routine use of antibiotics.
3   Eating locally produced organic vegetables whenever possible.
4   Segregating waste streams/recycling.

For six years, we have invited staff to nominate their peer co-workers for a chance to be recognized for their extraordinary contributions to personal and environmental health, culminating in 25 awards per year. Our system CEO and President preside over the awards event to honor the winners, so that embracing sustainability is truly a badge of honor at our organization.

What are the tangible implications of your annual environmental stewardship awards program?

The awards program has a positive influence on morale, a ripple effect. The people who win are engaged, and they bring others to the awards event, so awareness is spread both inside and outside of the organization. We publicize this event through videos and other online media and play the videos in board meetings. We are creating goodwill and awareness from the staff to senior leadership up to board level. Participation in the awards program has been steady since its inception. When it comes to how we measure our employee engagement, the environmental stewardship awards are one of our differentiators. We are building a culture of conservation and accountability. Employees are recognized for reducing energy and waste and thus become role models to others as well as stewards for our organization, which has a ripple effect.

What are your greatest challenges around sustainability at Advocate Health Care?

Some of the most basic minimization efforts, like recycling, remain a challenge. In common areas, we see a lot of litter. By and large, people are more careless in common areas; contamination of recycling streams is common. When it's not convenient, people will discard items in the closest bin, and do not think to recycle. Changing behavior is another challenge, especially the behavior of visitors. Behind the scenes, people are geared to be sustainable, but when the public comes into Advocate, these practices do not carry over and we do not have control over their compliance.

Introducing composting is another challenge for us, as landfills remain relatively cheap and composting is a more expensive disposal method. It's hard to make a business case for composting. Until composting becomes a law or the cost of solid waste disposal increases, composting will be hard to execute. I would love to see public policies requiring hospitals to compost or bio-digest food waste, which diverts waste from landfills.

What tools or resources have you found most helpful in your sustainability journey?

The peer involvement of the Healthier Hospitals has helped our program mature as well as identify and realign our priorities. Peer organizations have given us metrics by which to measure and benchmark our success.

The US Green Building Council has been a useful resource for our construction projects. More resources are coming around the bend pertaining to how healthcare can respond to climate change – we are starting to get into this dialogue. We are learning, too, from other industries (e.g., the hospitality sector) and the sustainability progress they have already made.

Any advice for latecomers to sustainability?

Get on board as fast as you can! You don't have to know everything. A lot has been done already by the trailblazers in the sector, so you are not in the dark. You can follow leaders who have proven the benefits of sustainability and who have built a culture of conservation. There is no excuse not to get on board anymore. Sustainability is a no-cost initiative. It's healthy for your bottom line. You are losing out if you are not implementing sustainability initiatives. Sustainability will attract employees to work at your organization, especially younger generations.

Practice Greenhealth

Practice Greenhealth is the nation's leading membership and networking organization for institutions in the healthcare community that have made a commitment to sustainable, eco-friendly practices. Members include hospitals, healthcare systems, businesses and other stakeholders engaged in the greening of healthcare to improve the health of patients, staff and the environment.[7]

Practice Greenhealth is a not-for-profit organization that provides visioning support for hospital staff through educational webinars, consulting and other training materials.

Health Care Without Harm

Health Care Without Harm is an international coalition of hospitals and healthcare systems, medical professionals, community groups, health-affected constituencies, labor unions, environmental and environmental health organizations and religious groups.[8]

Healthier Hospitals Initiative

The Healthier Hospitals Initiative (HHI) is a US-based coalition of major health systems and three key non-profits that are committed to improving sustainability and safety across the healthcare sector. The HHI agenda is based on the premise that a coordinated sector-wide approach to how we design, build and operate hospitals can improve patient outcomes and workplace safety, prevent illnesses, create extraordinary environmental benefits, and save billions of dollars. HHI utilizes research data on environmental sustainability and community health to delineate a prioritized sustainability roadmap, a method for moving from good ideas implemented by individual facilities to a comprehensive sector-wide adoption of sustainable practices and cost reduction.[9]

Global Green and Healthy Hospitals Network

Hospitals, health systems and health organizations representing the interests of more than 3,500 hospitals from six continents have come together to form a global network dedicated to reducing their ecological footprint and promoting public environmental health.[10]

Environmental Protection Agency

The mission of EPA is to protect human health and the environment [...]. When Congress writes an environmental law, [EPA] implement[s] it by writing regulations. Often, [EPA] set[s] national standards that states and tribes enforce through their own regulations. If they fail to meet the national standards, [EPA] can help them. [EPA] also enforce[s] [its] regulations, and help[s] companies understand the requirements.[11]

EPA has resources dedicated to the healthcare business sector, notably Energy Star, "a joint program of the US Environmental Protection Agency and the US Department of Energy helping us all save money and protect the environment through energy efficient products and practices."[12]

Department of Energy

The US Department of Energy (DOE) invites healthcare organizations to join the Hospital Energy Alliance (HEA). HEA brings together leading hospitals and national associations in a strategic alliance designed to improve energy efficiency and reduce greenhouse gas emissions of healthcare systems throughout the country. By leveraging access to advanced technologies emerging from the national laboratories, HEA members are creating a national forum for the industry to share evidence-based technology solutions and influence the energy performance of medical equipment and systems.[13]

United States Green Building Council

"The US Green Building Council (USGBC) is a 501(c)(3) non-profit community of leaders working to make green buildings available to everyone within a generation."[14] USGBC provides "green" building tools for the architecture, design and construction community through a credit system called Leadership in Energy and Environmental Design (LEED).

SUPPORTING ORGANIZATIONS

The following list is a snapshot of the health service professional organizations that recognize the importance of sustainability and share sustainability information with their members:

- American College of Healthcare Executives (ACHE)
- American Hospital Association (AHA)
- American Nurses Association (ANA)
- American Public Health Association (APHA)
- American Society for Healthcare Engineering (ASHE)
- Association for the Advancement of Sustainability in Higher Education (AASHE)
- Association of University Programs in Health Administration (AUPHA)
- The Center for Health Design (HCD)

Sustainability: a means to an end or an end in itself?

One of the primary choices a leader must make is whether to frame sustainability as a strategic imperative in its own right or as a means of accomplishing the institution's operational objectives. In recent years it has become increasingly clear that embracing sustainability can be an economically successful strategy – an effective means of accomplishing cost control because of the reduction in energy utilization, reduction in waste stream and environmentally preferable purchasing, for instance.[15]

At the same time the hospital's societal imperative is to serve as a community leader in the campaign to improve health, in part through performance as a sustainable organization. Boards of trustees and community advocates are looking past the role that sustainability can play in achieving operational goals; they are pressing for healthcare to take on the mantle of leadership as a champion of sustainability, both in terms of advocacy and of a model practitioner.

A key task of the healthcare leader who embraces sustainability is to define what excellence in sustainable performance looks like.[16] To begin with, this means developing sustainability mission and vision statements. It also means engaging in sustainability through leadership and program innovation; it requires a long-term commitment to investing resources and personal prestige in sustainability to create a high-performing entity.[17] This concept is no different from the process of establishing one's organization as a top decile performer in safety or finance.

A 2011 survey of 500 business leaders from The Business of a Better World demonstrates the confluence of sustainability and business success. Eighty-four percent of business leaders surveyed stated that their domestic company will adopt sustainability as part of its core business strategy over the next five years. Respondents expressed optimism that global companies will adopt sustainability within their core business strategies and functions over the next five years, indicating widespread concern about the environment and the strategic positioning of business interests.[18] There has been a shift since that time to focus overtly on sustainability (Makower, 2017).

The relationship between sustainability and climate change has a strong influence on strategic thinking in the business community. A PricewaterhouseCoopers (PWC) report highlights that 75 percent of CEOS agree "that satisfying societal needs (beyond those of investors, customers and employees) and protecting the interests of future generations is important" (PWC, 2017). In addition, "84% of respondents believe that business can pursue its self-interest while doing good work for society" (PWC, 2017). This suggests that a significant majority of surveyed corporate executives view climate change issues as a consideration of overall business strategy in terms of opportunities and risks. A McKinsey and Company study also had similar findings. Specifically, nearly 70 percent of executives surveyed view climate change influences as key to managing corporate brand, product development, investment planning, purchasing and supply management. Further, approximately one-third of respondents indicate that climate change issues trump most other global trends.[19]

It is important for business leaders to communicate the goals of sustainability initiatives. Within healthcare, people throughout the organization – doctors, volunteers and employees – need to understand at the grassroots level how their interests and personal goals connect to the organization's sustainability objectives. In the final analysis, there must be an alignment between the interests of the hospital's stakeholders and the sustainability program. Stakeholders want to be part of something greater than themselves and sustainability can be the platform.[20]

FRED: [*talking to himself*] A healthcare organization ought to be a role model for reducing waste, conserving energy, increasing community goodwill and brand equity, employee engagement and physician engagement. The process of managing every one of those critical performance criteria has to be about enhancing health. That means providing healthcare while having as few negative side effects as possible, doing so efficiently, consuming as few resources as we can. Consider what healthcare would be like if we applied the principle of sustainability to every major dimension of leadership. [*Knock on the door. Ms. Crumpmuffet, Fred's executive assistant, sticks her head in to let Fred know that his next appointment is ready. It's Felipe, Vice President of Materials Management.*]

FRED: Good afternoon, Felipe.

FELIPE: Good afternoon, Fred.

FRED: Did Denise connect with you regarding our sustainability conversation?

FELIPE: Yes, she did.

FRED: Very good. Well then let's dive right in. I think there is some element of a moral imperative here when people think about sustainability. There's not only a connection with the environmental goals and outcomes, but a call to "join the movement." That should be our message to our employees and staff, supported by suggestions of how to apply it in their daily activities.

FELIPE: That makes sense. I think one of the things I've learned while educating employees about supply chain management is that it's not about trying to give people a lecture or information. In my colleagues' eyes, it's less about our grand goal for the impact on the environment, and more about "What's in it for me?" and "What can I do that's going to make a difference?"

FRED: Of course, of course. Our challenge is to make sustainability something that every employee, at every level, feels connected to. It's about creating a culture.

Growing interest

The healthcare industry has seen significant growth in sustainability. Studies show that healthcare is not the only field in which sustainability is of growing interest. An analysis of businesses in 2010 shows that, even during economically challenging times, many companies increased investments in their sustainability activities and made bold new sustainability commitments. The 2017 *State of Green Business Report* shows a dramatic shift occurring in mainstream business: companies are thinking bigger picture and longer term about sustainability than they did in the past (Makower, 2017). In fact, they are focusing on outright positive and sustainable strategies rather than simply being focused on inadvertent negative consequences (Makower, 2017). This is a continual growth from the 2010 report. According to

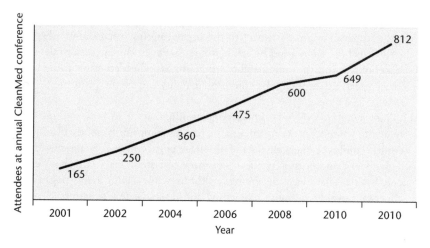

FIGURE 2.2 Growth of interest in healthcare sustainability

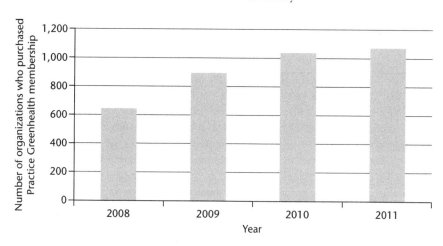

FIGURE 2.3 Growth of sustainability membership organizations

Farient Advisors, approximately 10 percent of Standard & Poor's 100 are currently factoring environmental issues into their incentive compensation plans, either as strategic objectives or as quantifiable goals.[21] *Newsweek* further supported the trend by releasing its third annual US and Global Green Rankings, which are beginning to influence corporate behavior and disclosure.[22]

SUSTAINABILITY: GENERAL FACTS AND FIGURES

- The Top 10 Best Colleges in America ranked by US News and World Report have sustainability training programs.[23]
- A recent survey found that commitment to environmental, social and governance issues is becoming exceptionally strong: 93 percent of CEOs

see sustainability as critical to their company's success. Sustainability is a strategic priority for executives around the world.[24]

- Eighty-one percent of Fortune 500 companies now issue a corporate social responsibility report. Global Reporting Initiative-based reports are at an all-time high (GAI, 2015).
- The number of companies reporting to the Carbon Disclosure Project (CDP) has increased tenfold in seven years (from 350 in 2003 to 3,050 in 2011).[25] Currently, 80 percent of Fortune 500 companies are reporting to the CDP (Shiraishi, 2014).
- According to a report produced by Ernst & Young, "In 2010, resolutions focusing on social and environmental issues made up the largest portion of all shareholder proposals. That trend is expected to continue this year: we estimate that half of all shareholder resolutions in 2011 will center on social and environmental issues" (Ernst & Young, 2011). This trend has continued, with data from 2014 to 2016 highlighted in Figure 4.5.

A growing field

The growing priority of sustainability in business is consistent with trends in higher education. A 2010 survey by the Association for Advancement of Sustainability in Higher Education (AASHE) documents that 260 of the 421 universities surveyed have sustainability positions created since 2008.[26] According to the AASHE survey (2010), just over 40 percent of sustainability-related leaders had sustainability work experience of five years or less, excluding internships and activism,[27] which is consistent with the theory that many people choosing to work in the field of sustainability represent the future workforce. In addition, a 2014 study highlighted a jump

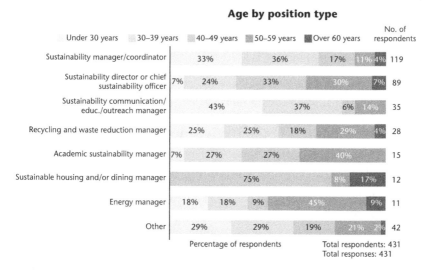

FIGURE 2.4   Sustainability position type by age

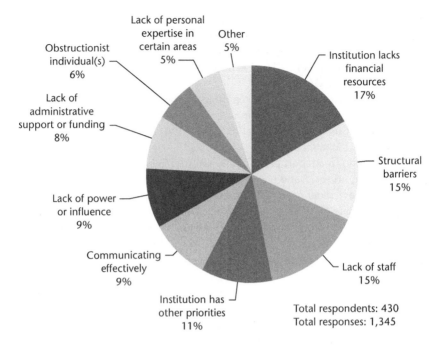

FIGURE 2.5 Biggest challenges of sustainability positions

from 30 percent to 43 percent in executives that seek to align sustainability with their overall business goals and missions.

The AASHE survey also described the challenges which sustainability leaders face, including:

- institutional lack of financial resources (17%)
- structural barriers (15%)
- lack of staff (15%)
- institution has other priorities (11%)
- effective communication (9%)
- lack of power or influence (9%)
- lack of administrative support or funding (8%)
- obstructive individual(s) (6%)
- lack of personal expertise in certain areas (5%).

A sustained investment in sustainability: does it make sense?

Asking why leaders invest in sustainability is like asking why leaders invest in brand. It is not about immediate, short-term returns. In *The Seven Habits of Highly Effective People*, Stephen Covey writes that great organizations invest in third-quadrant priorities. There are four quadrants of leadership priorities based on urgency and importance.[28] Focusing on sustainability is a third-quadrant investment along with building brand and satisfying customers, training employees and investing in the next care delivery model.[29] When economic times are hard, successful organizations do not stop research and development; nor do they stop investing in employee engagement. Similarly, tough economic times provide no excuse for ignoring sustainability imperatives.

A Pew Charitable Trust study documents that "green-collar jobs"[30] in the USA grew by 9.1 per cent between 1998 and 2007, approximately 2.5 times faster than job growth in the overall economy.[31] While the task of defining and quantifying green jobs remains a challenge, the Brookings Institution's report *Sizing the Clean Economy* estimates that nearly 2.7 million people worked for the "clean economy" in 2010.[32] To encourage momentum in the clean economy, President Obama signed an appropriations act in December 2009 that provided $611.4 million to the Bureau of Labor Statistics for the 2010 fiscal year, targeted specifically to stimulate the production of green jobs, industries and occupations.[33] According to *Fortune* magazine, the renewable energy sector is creating jobs 12 times faster than the rest of the economy (Samuelson, 2017). In other words, society is increasingly focused on sustainability across market sectors. Some may question how much changes in government will influence these trends. While there is no glass ball, marketplace demands trend in favor of sustainability, regardless of political leadership.

[*Denise, Felipe and Fred engage in conversation.*]

FELIPE: We have data from other organizations showing that there's a strong financial benefit that can result from a long-term approach to investing in a focused sustainability program. For instance, Inova Health System saved $1 million this year by reprocessing medical devices.[34]

FRED: Sure, that's impressive, but we're a stand-alone community hospital and Inova is a system of hospitals.

DENISE: Well, there are plenty of examples of community hospitals that are performing well and saving money through their sustainability programs. Gundersen Lutheran is an impressive example of a community hospital that excels at energy management. They partner with City Brewery in La Crosse, Wisconsin to generate three million kilowatt hours per year, turning discharged biogas waste from City Brewery into electricity for the hospital. I contacted their sustainability champion and learned that Gundersen Lutheran intends to lower its energy consumption costs by 20 percent annually, eventually becoming 100 percent energy independent by 2014.[35] They are an energy-independent healthcare system now, and they're capturing biogas waste and reducing demand on the power grid.

FRED: Wow, officially off the grid! Who would've thought a beer distributor would be a sustainability partner? It's ironic that one of my early memories of discussing sustainability is someone telling me, "There's no cash in the trash." The person's point was that if you're only practicing sustainability to save money, that's an inadequate justification. I'm realizing that sustainability does go beyond moral obligation. While we're unlikely to get rich through sustainability initiatives, sounds like sustainability can be fiscally rewarding in some instances. Today, based on what you're telling me about sustainability programs in other systems, there are sustainable practices we could engage in that have positive environmental and financial returns. If we're going to get serious about practicing sustainability at Memorial Hospital, we need a sustainability director. We need someone who wakes up every day and thinks about how he or she will drive sustainability throughout the hospital.[36]

FIGURE 2.6 Engaged leadership: an overview

DENISE: Once our leaders commit to being sustainability-minded, we'll need a steering committee. Perhaps a hub-and-spoke model, to connect our sustainability program to each dimension of our organization. Each of those spokes usually has a leader.[37]

FELIPE: Why are you suggesting the hub-and-spoke model as opposed to any other organizational framework?

DENISE: One of the reasons to employ the hub–and–spoke model is because there's a grassroots level of interest with sustainability. We hear that people want to be engaged in our sustainability program and they offer a variety of thoughtful ideas. We should connect with the frontline and connect leadership with the good ideas that employees are sharing. There are a lot of people in this organization who care about sustainability. We need to give them vehicles to participate, to connect with the sustainability mission and cultural change projects. I think it will help immensely to have a sustainability director – someone to facilitate sustainability progress across the organization.

MANAGEMENT TIP

It is a misconception to think that sustainability is mostly a top-down process. Success, whether in sustainability or other domains, is often driven by how talented the performer is and how well he or she performs for the judges. In short, success requires picking projects carefully, connecting with natural leaders, and engaging the broader population of employees, physicians and the community. This combination results in leaders and employees engaging in the hospital's work of sustainability with little further encouragement other than connecting it to the core culture of the hospital.

Implementation: build sustainability into an organization

Sustainability impacts every dimension of a hospital, from purchasing decisions, to nursing care, to facility design and maintenance, food services, operating room regimens, strategic planning, safety standards, quality efforts and community benefit.[38] Data from Practice Greenhealth reports strongly suggest active cross-departmental Green Teams in order to achieve successful sustainability programs.[39] Departmental representation on Green Teams includes:

- Administration
- Communication/marketing
- Engineering
- Environmental health and safety
- Environmental services
- Facilities
- Food services
- Infection control

- Materials management
- Nursing
- Nutrition
- Physician
- Risk management
- Safety.

Research further suggests that there is a trend toward identifying a sustainability owner for the hospital whose sole focus is to wake up every day with the intent of promoting sustainability.[40] Healthcare institutions that successfully integrate sustainability within the organizational culture often fulfill the following characteristics. They:

1 Have an environmental commitment statement.
2 Have a written plan for environmental management of the facility.
3 Have a Green Team.
4 Have a designated sustainability director.
5 Have someone on staff who is responsible for sustainability within their job description (but not by title).
6 Track their sustainability improvement initiatives in the Joint Commission structure.
7 Provide new employee orientation on sustainability initiatives.
8 Provide annual training on sustainability initiatives.
9 Engage clinicians, including nurses and physicians, in their sustainability programs.

Organizations that embrace these fundamental sustainability program elements are most likely to improve the sustainability cultures of their organizations. Simply put, the emerging role of the sustainability director in healthcare is increasingly important.

Practical application: hardwiring environmental stewardship at Advocate Health Care

The imperative

- Harnessing the interest and ideas of passionate individuals.
- Organizing and integrating a sustainability infrastructure.
- Improving the patient experience through engaged, happy employees.
- Creating a culture of sustainability.
- Capturing and maximizing dollar savings opportunities.
- Identifying the role of each employee as it relates to sustainability.
- Aligning employee values and creating a culture of similar/system thinking.
- Maximizing the organization's impact upon public health.

The initiative

Advocate Health Care takes a multi-pronged approach to hardwiring environmental stewardship into the core of its operations. This engagement program targets both frontline employees and leadership. It focuses on helping these groups understand both the *why* and the *how* of sustainability, providing engagement mechanisms that allow every stakeholder to take part in achieving the organization's broader environmental stewardship goals.

The role of Advocate's dedicated sustainability position is to help the organization and community understand what opportunities are available and to facilitate their implementation when possible.

Advocate incentivizes employees to engage in sustainability by appealing to their intrinsic desire to do good and by offering ways to be involved. Examples of such initiatives include:

- Green Teams at each facility.
- System-level working groups comprising Green Team leaders to identify priorities and action items.
- Annual computer-based environmental stewardship tutorials.
- Green Advocates Ambassador Program, a self-sustaining program that incentivizes the creation of sustainable workspaces and rewards entire departments for their participation in "greening" their workspaces.
- Internal awards program for employees engaged in environmental stewardship.

Advocate's efforts in executive engagement focus on engaging top leadership teams to create a setting where participation opportunities are available to every employee. Some of their mechanisms for achieving this goal include the following:

- Sustainable Operations Council comprising VP-level-and-above leaders from every site that has a Green Team, as well as key representatives from every department.
- Sustainable Operations Index comprising specific environmental goals for the organization, which each person on the System Committee and Sustainable Operations Council carries as part of their own set of goals, and which are rated in performance reviews.
- Quarterly Environmental Stewardship Dashboard summarizing sustainability performance is publicly visible, and distributed to community and council members and all hospital leadership personnel.

The metrics

Although the organizational integration of a sustainability program is difficult to measure quantitatively, certain metrics assist in gauging progress in this endeavor:

- Levels of participation in employee engagement events and training programs.
- Responses to sustainability questions on employee engagement surveys.

- Performance across a broad suite of Sustainability Dashboard measures, such as recycling rate, regulated medical waste per adjusted patient day (APD), construction and demolition recycling percentage, kBtu (kilo-British thermal unit) consumption (with the goal of percentage reduction), greenhouse gas emissions, dollars saved in electricity/energy use, reprocessing savings, paper reduction, etc.
- Overall progress toward achieving organizational sustainability goals identified by the Sustainable Operations Council.

Lessons learned

- To seek help from leadership in driving these efforts from the top down, it is necessary first to translate the concepts into leadership's language and to establish accountability for achieving outcomes.
- Support and passion from key leadership are critical dimensions of achieving the successful adoption of sustainability programs: work to achieve a "cascading" effect from the top down. Waste less time with personnel who are not on board or not passionate about the sustainability mission. Instead, tap into the energized individuals. They will make things happen. Do not waste your time trying to convince people who do not want to be convinced, especially during the early stages of implementation.
- It is crucial to harness the passion that exists in all staff, which is easiest to achieve when organizational incentives align with employees' professional and personal interests. In the healthcare setting, emphasizing the connection of sustainability and wellness is an alignment worth considering.
- Hiring a person to organize the sustainability efforts and translate between stakeholders will facilitate the implementation of sustainability projects. The individual filling this role will be most effective if she or he has full access to the organization's resources and reports to as high a rank as possible within the organization.
- Getting buy-in from certain departments may be challenging. Each department's and individual's incentives may be different and sometimes it will be tough to find a champion. Disconnects will exist. Helping each department and individual understand *why* sustainability matters will help cement shared understanding about the purpose and value of sustainability.

About the organization

Recognized as one of the nation's top ten health systems, Advocate Health Care is the largest integrated healthcare system in the State of Illinois. The faith-based, not-for-profit health system offers more than 250 sites of care, including ten acute-care hospitals and two integrated children's hospitals.

Sources

Advisory Board (2017). CMS: US health care spending to reach nearly 20% of GDP by 2025. February 16. Available at www.advisory.com/daily-briefing/2017/02/16/spending-growth (accessed August 10, 2017).

Governance and Accountability Institute Inc. (2015). FLASH REPORT: 81% of S&P 500 companies published sustainability reports in 2015. Available at www.ga-institute.com/press-releases/article/flash-report-eighty-one-percent-81-of-the-sp-500-index-companies-published-corporate-sustainabi.html (accessed August 10, 2017).

Healthcare Without Harm (n.d.). *The Global Green and Healthy Hospitals Agenda.* Available at https://noharm-global.org/issues/global/global-green-and-healthy-hospitals-agenda (accessed August 10, 2017).

Makower, J. (2017). The State of Green Business. January 31. Available at www.greenbiz.com/article/state-green-business-2017 (accessed August 10, 2017).

PricewaterhouseCoopers. (n.d.). Sustainability – business success beyond the short term. Available at www.pwc.com/gx/en/services/sustainability/ceo-views-sustainability-perspective.html (accessed August 10, 2017).

Samuelson, K. (2017). Renewable energy industry creates jobs 12 times faster than rest of U.S. January 27. Available at http://fortune.com/2017/01/27/solar-wind-renewable-jobs/ (accessed August 10, 2017).

Shiraishi, A. (2014). 6 reasons your company should report to CDP. June 6. Available at www.greenbiz.com/blog/2014/06/06/6-reasons-your-company-should-report-cdp (accessed August 10, 2017).

www.advocatehealth.com/body_nonav.cfm?id=1427.

Larsen, Mary. Personal interview, May 1, 2012.

Leadership imperative

People who practice sustainability for moral reasons will do so regardless of the financial costs. There additionally exists a business case for investing in sustainability as a practical, "here-and-now" component of healthcare leadership. Leadership has an obligation to build sustainability into organizational mission and vision because of healthcare's commitment to preserve and improve the community's health. Leaders cannot stop at mission and vision; sustainability must also be incorporated into operational performance if its full potential is to be realized.

Leadership dialogue

1   Why do you think sustainability should or should not be a core mission and responsibility of healthcare leaders and organizations?

2   Executive team engagement is essential to drive core values throughout the organization, but engaging executive leadership may be particularly challenging in the case of sustainability, which is not a traditional leadership priority. What are strategic approaches to engage those leaders who may be most reluctant to get on board?

3   Is having high technical knowledge or sincere passion for sustainability the most important quality in an executive sustainability champion? Why?
4   Covey's "third quadrant" focuses on priorities that are important but not urgent. Why is it – or why isn't it – appropriate to cite sustainability as an example of third-quadrant thinking?

Syllabus integration

Overview

Healthcare professionals with an understanding of sustainable practices share a unique vantage point of challenges and solutions that cut across business; they are positioned to identify problems and opportunities and to broker information. This lesson incorporates tools used to measure environmental health system impacts, focusing on innovative solutions that add business value and contribute to an organization's overall strategic business success.

Objectives

1   To facilitate an oral presentation of homework from the previous class.
2   To identify critical interactions between economic, social and sustainable development issues that are important to business activities.
3   To examine the planning, development and implementation considerations taken into account when evaluating sustainability drivers and programs.
4   To learn how to advance business performance through metrics program implementation.
5   To evaluate business benefits, costs and best practices by systematically addressing economic development, social equity and environmental issues.
6   To review sustainability job descriptions.

Topics

1   Plan–do–check–act (PDCA) life cycle
2   Strategic development planner
3   Three safeties
4   Future sustainability employment opportunities
5   Performance reporting

Homework

Healthier Hospitals. *Engaged Leadership*. Available at http://healthierhospitals.org/hhi-challenges/engaged-leadership.
Imagine you are the CEO of a healthcare organization that is newly committed to sustainability practices. Based on classroom conversation and examples, write an organizational statement of commitment to sustainability to a public community audience. Record the statement in your career journal.

Practice Greenhealth (2012). *Sample Job Description: Healthcare Sustainability Director.* Reston, VA: Practice Greenhealth. Available at https://practicegreenhealth. org/pubs/toolkit/greenteam/SampleJDSustainabilityDirector.pdf.

Draft a healthcare sustainability administrative job position in your career journal. Include a job title, position description (required skills and experience, core responsibilities), compensation (salary and benefits) and location on the organizational hierarchy.

Additional resources

Environmental Health. *Tools for Nurses*, United American Nurses, AFL-CIO.

Healthier Hospitals. *Engaged Leadership Resources.* Available at http://healthier hospitals.org/hhi-challenges/engaged-leadership#resources.

Sustainable Development Unit. *What is Sustainable Health.* Available at www. sduhealth.org.uk/policy-strategy/what-is-sustainable-health.aspx.

Notes

1 Mathur, Y. (2009). Health care reform explained. *Business Today.* Available at www. businesstoday.org/magazine/its-always-christmas-washington/health-care-reform-explained.
2 Davis, K., Schoen, C. and Stremikis, K. (2010). *Mirror, Mirror on the Wall. How the Performance of the US Health Care System Compares Internationally.* Washington, DC: Commonwealth Fund. Available at www.commonwealthfund.org/publications/fund-reports/2014/jun/mirror-mirror.
3 The Health Care Marketplace Project (2011). Health care spending in the United States and selected OECD countries. Kaiser Family Foundation. Available at www.kff.org/health-costs/issue-brief/snapshots-health-care-spending-in-the-united-states-selected-oecd-countries/.
4 Advisory Board. (2017). CMS: US health care spending to reach nearly 20% of GDP by 2025. February 16. Available at www.advisory.com/daily-briefing/2017/02/16/spending-growth (accessed August 10, 2017).
5 Health Care Without Harm (n.d.). The campaign for environmentally responsible health care. Available at http://infohouse.p2ric.org/ref/16/15197.pdf.
6 Elkington, J. (n.d.). Enter the triple bottom line. Available at www.johnelkington.com/archive/TBL-elkington-chapter.pdf.
7 http://practicegreenhealth.org.
8 www.hcwh.org/.
9 http://healthierhospitals.org.
10 http://greenhospitals.net.
11 www.epa.gov.
12 www.energystar.gov/index.cfm?c=about.ab_index.
13 EERE Information Center (2011). *Commercial Building Energy Alliances: Making the Business Case for Energy Efficiency.* Available at www1.eere.energy.gov/buildings/publications/pdfs/alliances/commercial_building_energy_alliances_fact_sheet.pdf.
14 www.usgbc.org.
15 Cram, P., Bayman, L., Popescu, L., Vaughan-Sarrazin, M., Cai, X. and Rosenthal, G. (2010). Uncompensated care provided by for-profit, not-for-profit, and government owned hospitals. *BMC Health Services Research* 10(90): 1–13. Available at www.ncbi.nlm.nih.gov/pmc/articles/PMC2907758/.
16 Berchicci, L. and Bodewes, W. (2005). Bridging environmental issues with new product development. *Business Strategy and the Environment* 14: 272–85.

17 Bansal, P. and Roth, K. (2000). Why companies go green: a model of ecological responsiveness. *Academy of Management Journal* 43(4): 717–48.
18 McKinsey & Company (2008). How companies think about climate change: a McKinsey global survey. *McKinsey Quarterly*, February. Available at www.mckinseyquarterly.com/How_companies_think_about_climate_change_A_McKinsey_Global_Survey_2099.
19 Colburn, M. (2011). Overcoming cynicism: Lord Hastings' call to action at Net Impact. *GreenBiz*, November 2. Available at www.greenbiz.com/blog/2011/11/02/overcoming-cynicism-lord-hastings-call-action-net-impact.
20 McKinsey & Company (2008).
21 Ferracone, R. (2011). The role of environmental sustainability in executive compensation. *Forbes*, April 26. Available at www.forbes.com/sites/robinferracone/2011/04/26/the-role-of-environmental-sustainability-in-executive-compensation.
22 Buechner, S. and Ellis, M. (2011). 3 ways to improve green business rankings. *GreenBiz*, November 3. Available at www.greenbiz.com/blog/2011/11/03/3-ways-improve-green-business-rankings.
23 Myslymi, E. (2016). U.S. News & World Report Releases 2017 Best Colleges Rankings, September 13. Available at www.usnews.com/info/blogs/press-room/articles/2016-09-13/us-news-releases-2017-best-colleges-rankings (accessed August 11, 2017).
24 Lacy, P. (2010). On the verge of a sustainability "tipping point". *GreenBiz*, June 24. Available at www.greenbiz.com/blog/2010/06/24/verge-new-era-sustainability.
25 Baier, P. (2011). How leading firms make their sustainability reports stand out. *GreenBiz*, May 3. Available at www.greenbiz.com/blog/2011/05/03/how-leading-firms-make-their-sustainability-reports-stand-out.
26 AASHE (2010). *Higher Education Sustainability Staffing Survey*. Denver: Association for the Advancement of Sustainability in Higher Education. Available at http://o.b5z.net/i/u/10069179/f/AASHE_Higher_Education_Sustainability_Staffing_Survey.pdf.
27 Ibid.
28 Covey, S.R. (1989). *The Seven Habits of Highly Effective People*. New York: Simon & Schuster.
29 Ibid.
30 US Bureau of Labor Statistics (2010). The 2010 President's Budget for the Bureau of Labor Statistics. Bureau of Labor Statistics. Available at www.bls.gov/bls/budget2010.htm.
31 Galbraith, K. (2009). Study cites strong green job growth. *New York Times*, June 10. Available at http://green.blogs.nytimes.com/2009/06/10/study-cites-strong-green-job-growth.
32 Muro, M., Rothwell, J. and Saha, D. (2011). Sizing the clean economy: a national and regional green jobs assessment. *Brookings Report*, July 13. www.brookings.edu/research/sizing-the-clean-economy-a-national-and-regional-green-jobs-assessment/.
33 US Bureau of Labor Statistics (2010).
34 Inova Health System (n.d.). Inova's environmental leadership. Available at www.inova.org/about-inova/sustainability/history.jsp.
35 Anon. (n.d.). Gundersen Lutheran teams with brewery on unique "green" endeavor, Gundersen Lutheran. Available at www.gundluth.org/?id=3069&sid=1.
36 Brown, J. (2010). *Benchmarking Sustainability in Health Care Awards Benchmark Report*. Reston, VA: Practice Greenhealth. Available at https://practicegreenhealth.org/tools-resources/sustainability-benchmark-report-0.
37 Griffith, J.R. and White, K.R. (2007). *The Well-managed Healthcare Organization*, 6th edn. Chicago, IL: Health Administration Press.
38 Brown (2010).
39 Ibid.
40 Ibid.

3

SALUTING THE CAPTAIN OF THE SHIP

Process engagement + health information technology

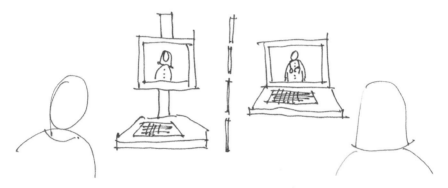

FIGURE 3.1 Engaging physicians with health information technology

Executive summary

A physician's concern is about the health of his or her patients. Physicians understand the concept of the "health of the community," but prioritize the welfare of their individual patients more highly.[1] Physicians are aware that their patients expect a safe, environmentally friendly healthcare setting. This chapter explores physicians' engagement through the lens of sustainability in a delivery setting with health information technology.

Storyline

Director of Marketing and Communication, Fiona O'Malley, is working with Dr. Hedge, Chief Medical Officer, on the communication strategy for updates to the electronic health record called Epic.

FIONA: Dr. Hedge, we are working on the newsletter about updates to Epic. I've added the graphics. What are your thoughts? [*Fiona hands the drafted newsletter to Dr. Hedge for his approval.*]

DR. HEDGE: Am I looking at the right document? I'm seeing a picture of trees and a forest. What do these images have to do with Epic?

FIONA: Our Sustainability Director told me about the environmental impacts of the new electronic medical record. We'll significantly reduce the amount of paper that's being used with the Epic updates. And you know, the way she explained it helped me understand that reducing paper usage is ultimately about patient health. Reducing paper not only saves trees, but leads to a significant reduction in water use and chemical and greenhouse gases that are used in paper production. Changing our documentation process will have an especially strong impact upon our community because we have that pulp mill just outside of town. Think about the air quality around here and what that means for our pulmonary patients.

DR. HEDGE: I never thought of it like that. My son's nanny works at the pulp mill part time and I've wondered about the health impacts of working there. I never thought about the connection to what we do here though [...]

FIONA: Dr. Hedge, you might consider meeting with Amalie Dubois, our Sustainability Director. I bet she could tell you more about what we're doing in terms of sustainable practices, and what we could be doing. But in the meantime I need to get this story out, so what are your thoughts on the newsletter draft?

DR. HEDGE: Well Fiona, I like where you're going with this, but let's take it a step further. Could we print the newsletter on recycled paper, or perhaps distribute the newsletter electronically?

If IT's important to the patient, IT's important to the physician

Embracing sustainability may add value to the hospital's relationship with the physician if it adds value to the physician–patient relationship.[2,3] To the degree that patients care about sustainability, they may care about the physician's regard or disregard for sustainability. It is felt by many observers that sustainability will increasingly be part of how physicians relate to, and are evaluated by, their patients (Nayler and Appleby, 2012).

The physician, health information technology and sustainability

Mrs. Lobatnic enters the hospital to pay a visit to Dr. Goldberg. Once again, Mrs. Lobatnic has no physical ailments to address during the appointment; she just wants to talk. Today she wants to talk about her grandson's science fair project on sustainability. Dr. Goldberg listens to Mrs. Lobatnic, checks to ensure she's feeling okay, listens to a few stories, and then proceeds to the medical staff meeting. At the medical staff meeting, the Chief Technology Officer, Lucient Uguayo, explains to the physicians that he's been charged with integrating sustainability into Memorial Hospital's technology practices. Lucient communicates to the medical staff that sustainability integration will begin in Women's Services. This news is not received well by the physicians.

Physician: captain of the ship

To engage the team, it is essential to engage the leader. The idea of engaging the team without engaging the leader is an oxymoron. It is like an athletic team having no coach to assign positions or orchestrate plays. Physicians' support and participation are integral to achieving almost all priorities of any hospital. To build a culture of sustainability, the healthcare team must carry out "plays," similar to an athletic team. The quarterback of the team is the physician and the teammates are nurses, who are important contributors to and influencers of a hospital's culture.[4]

The physician's role

Physicians are respected members of the community who exert tremendous social influence. The physician's role, in the hospital and in the community, is important to the success of the sustainability effort.[5] It follows, then, that engaging physicians involves making clear how sustainability is important to physician-valued programs in the hospital. Physicians are, of course, also motivated by the economic success of their practice and their often symbiotic economic relationships with their admitting hospitals.[6] For example, the anesthesiologist will be interested in sustainability when the hospital leader demonstrates how the anesthesia program is, in some fashion, dependent upon sustainability. This alignment of economic interests between the physician's practice and the welfare and success of the hospital is more direct for employed versus independent community practice physicians, but it exists for both groups. The proportion of physicians who are hospital employees or contractors will influence the degree to which this alignment is apparent and leads to an increasing level of physician support for sustainability initiatives.

From the administrative perspective, physicians are the clearly dominant economic customer in the US market. If a customer assessing a car for purchase values safety in a car, then safety becomes a priority for anyone selling cars. Similarly, healthcare leaders pay close attention to physicians as customers and usually find that physician values and priorities drive leadership and organizational values. Of course, boards and executive leadership articulate values and contribute to organizational priorities, but it is typically within the "excess space" defined by physician interests as the primary customer of the hospital.

If the goal of the board and executive leadership is to build the organizational cultural value of sustainability, that path must include meaningful engagement with physicians. Therefore it is essential to bring physicians into the sustainability dialogue at the earliest possible point and to identify those physicians who will champion the cause and engage their colleagues around sustainability. It is particularly important for leaders to effectively explain why sustainability complements physicians' professional objectives and how it can generate resources for physicians' priorities.

Hospitals are concerned about their public image; that is, how they are perceived by the community. Physicians are visible and influential participants in the community, and are positioned to carry the hospital's sustainability message to constituencies who view physicians as credible voices to be respected in the public dialogue. How can the leader familiarize physicians with the concept and applications of sustainability?

Fred is having lunch in the physician's dining room with Dr. Hedge, Chief Medical Officer, and several other medical staff leaders.

DR. HEDGE: Fred, good afternoon. How's the sustainability initiative going?

FRED: I'm learning how complicated and challenging it will be to accomplish our sustainability vision. I'm concerned about the people on the team and about sustainability's impact upon the broader organization. I'm concerned that if we don't have a sustainability program, it will negatively impact the community's perception of the hospital and our priorities. I've got to worry about how the patients and physicians perceive sustainability and experience sustainability. I think about the cost structure of our sustainability program because it'll cost something to get the program up and running. Sustainability never came up in my professional education. I learned how to read a balance sheet. I learned how to counsel employees, how to hire and how to resolve conflict. I learned about process improvement and how to promote a culture of safety. That's a long answer to your brief question, but the point is that learning about sustainability is an educational process for all of us. I'm not well prepared.

DR. RUPEE: [*Clinical Department Chair*] I've never thought about sustainability before and we've never talked about it before as a medical staff. I'm a department leader and I realize that if we're going to succeed, I've got to develop my people in this area.

FRED: Sustainability is an emerging competence that we should address using the same processes we used to address quality improvement and leadership development. The first challenge is to reach the folks who are current healthcare professionals. The second step is to see to it that people don't come out of education for medicine, nursing, health administration and the other health professions as poorly prepared in sustainability as we seem to be. The health administration programs are strong on medical care quality, finance and general management. The students get some exposure to sustainability issues in public health. But understanding sustainability is non-existent in most programs, at least the ones I've been exposed to. With your support and role as the Clinical Chair, Dr. Rupee, we will make great strides. Every physician, nurse, administrator and other staff member in our organization has to embrace sustainability if we're going to transform our culture. Those who are engaged are learning a lot about sustainability, me in particular.

DR. GEHRY: [*Internist*] I'm starting to see the link. If you're going to be a primary care physician, does that mean that you should know something about sustainability?

FRED: I believe sustainability is an indispensable part of the foundation of healthcare clinical and administrative leadership. I have to admit that I've become an advocate.

DR. HEDGE: I think we have several clinical leaders on board, though some are more enthusiastic than others. Physician engagement is challenging on most every subject, not only for sustainability. It's difficult to get physicians to recognize that sustainability – or any other goal indirectly connected to clinical care – is important to the organization. One challenge is that most of our staff physicians spend their time in their offices. While physical proximity isn't everything, it is generally easier to engage the hospital-based specialists.

DR. KUSEK: [*Chair of Obstetrics/Gynecology*] We depend a lot on our regular medical staff departmental meetings, but recognize their limitations. The informal networks are uneven. I am the first to admit that empowering and engaging physicians is the most challenging part of my job. It's not the knowledge or understanding; it's finding ways to actually connect with some of the busiest people I know.

FRED: When I was a young hospital administrator, I spent a lot of time in medical staff meetings. Nowadays there seems to be little physician interest in the hospital activities, especially in those specialties whose practices are largely office-based. I have to admit, I don't know how in the world we can get physicians interested and engaged in sustainability. Any ideas?

DR. HEDGE: You know, just yesterday I had a surgeon ask me, "What's the deal with this sustainability program I've been hearing about?" It was a great opportunity for me to connect with her and suggest that she engage with other physicians who've exhibited some interest in the sustainability program. I agreed to schedule a coffee break with them to dialogue about sustainability in more detail. If each of us pulls together two or three physicians for a sustainability conversation, we would move forward sustainability engagement.

Fred and the physicians agree to pursue Dr. Hedge's suggestion. As Dr. Hedge leaves the cafeteria, he pauses at the trash bin, checks the back of his food plate for the recycling symbol, and appropriately discards his plate in the recycling bin.

DR. RENATA: I have enough on my plate. I don't want to have anything to do with sustainability. I care about quality and this new patient care model. I don't want the sustainability thing.

LUCIENT UGUAYO: Don't you see, Dr. Renata, that there's a relationship between sustainability stewardship and stewardship of patient interests?

What does health information technology have to do with sustainability?

Health information technology (HIT) as an electronic medium largely replaces paper-based documentation in the clinical care process. Paper is processed using large quantities of water and energy. Paper is delivered to its intended location using trucks and gas. Hospitals with low levels of HIT implementation consume a lot of paper: paper forms, paper registration information, paper bills, paper records of all sorts.

On the other hand, the use of personal computers results in higher energy consumption.[7] There is a counter-argument about replacing paper with HIT that must also be considered due to the large amount of energy consumed by most HIT systems. While it is not simple to determine the "right" answer in terms of sustainability impact, Turley et al.'s study, among others, investigated the environmental impacts of electronic records and concluded that electronic health records have a positive net effect on the environment.[8]

LESS PAPER VERSUS PAPERLESS

Even hospitals that use electronic medical records maintain paper records due to state laws that require patient records to be stored for years (for example, patient records in Massachusetts must be stored for 25 years after the time they were last used).[9] "Iron Mountain, a data storage and protection company that claims to work with about a third of the nation's hospitals, estimates that the 6,000 hospitals in the US spend about $10 billion a year to store and manage 500 million patient records" and the cost to convert historical files is estimated to be "$1 billion for Johns Hopkins Hospital alone."[10]

Practical application: Kaiser Permanente's HealthConnect

The imperative

- Improving quality of care and service.
- Ensuring and continuously improving patient safety.
- Modernizing outdated systems and obsolete practices.
- Increasing access to latest treatments and protocols.
- Facilitating information-sharing.
- Emphasizing the organization's commitment to preventive medicine.

The initiative

HealthConnect is a comprehensive health information system that includes one of the most advanced electronic health records available. It is an innovative tool that enables Kaiser Permanente to deliver high-quality, safe and efficient care in new and unique ways. It facilitates communication between members and Kaiser Permanente's team of professionals to help make getting well and staying healthy as convenient as possible.

HealthConnect provides a website that facilitates e-connectivity between a patient and the healthcare team. It allows patients to view key components of their medical records and conduct clinical transactions online, providing patients with information to make knowledgeable decisions about their health. At the same time, HealthConnect allows physicians to contact patients electronically, order consultations, laboratory and other diagnostic work, send prescriptions directly to the pharmacy, provide health literature and set reminders for follow-up appointments.

Kaiser Permanente's HealthConnect was initiated in 2004 and completed in each of the organization's medical facilities in early 2010. It is the world's most widely used personal health record, enabling healthcare providers to connect with some 59 million patients through online portals. That volume of electronic communication accounted for 52 percent of Kaiser Permanente's total visits in 2015, which means that the health system now serves more patients online than in person.

The metrics

Kaiser Permanente found that a comprehensive electronic health record can increase consumer convenience and satisfaction and provider efficiency while maintaining clinical quality. Metrics Kaiser Permanente uses to measure the success of Health-Connect include:

- Rates of outpatient, urgent care and emergency department visits.
- External referrals.
- Scheduled telephone visits.
- Annual total office visit rates.
- Website traffic data.
- Use of particular website features.
- Repeat visits.
- Number of people registered to use the system.

Lessons learned

- It is important to have an economic model that aligns financial incentives with effective and efficient care, regardless of how care is delivered. Rewarding care strategies other than face-to-face visits will increase the incentives for providers.
- If face-to-face visits remain the gold standard for measuring quality, care standards will not reflect the preference of consumers for alternative, more convenient modes of care when they are appropriate or reinforce more efficient care delivery options.
- Collaboration among providers, combined with government support, will help advance widespread consumer adoption of virtual healthcare.
- Electronic health records may increase the time it takes to document patient visits, but since email messaging and scheduled phone visits consume less time than in-person visits, logic suggests that the efficiency gains offset any increases in documentation time. Temporary decreases in productivity of as much as 15 percent were reported during implementation.

About the organization

Founded in 1945, Kaiser Permanente is one of the nation's largest not-for-profit health plans, serving more than 110 million people, with headquarters in Oakland, California. It comprises the Kaiser Foundation Hospitals and their subsidiaries, the Kaiser Foundation Health Plan, Inc. and The Permanente Medical Groups.

Sources

Fast facts about Kaiser Permanente. Available at http://xnet.kp.org/newscenter/aboutkp/fastfacts.html.

Kaiser Permanente identifies key elements in successful healthcare information technology implementation. Available at http://xnet.kp.org/newscenter/press-releases/nat/2009/031009healthaffairs.html.

Kaiser Permanente HealthConnect® electronic health record. Available at http://xnetkp.org/newscenter/aboutkp/healthconnect/index.html.

Leadership imperative

To achieve personal ownership, any individual must answer the fundamental question: why should I care? The best link for engaging physicians in sustainability is the shared imperative around improving patient health. HIT is one example of a metaphor for improving patient health and reducing unintended harm to patients. Similarly, sustainability is an upstream solution many physicians can relate to population health.

While the health impacts may be the "hook" to get a physician's attention, his or her interest may be further engaged by the cost-saving opportunities: projects such as hibernating computers in non-clinical areas when not in use, telecommuting, or moving toward electronic pay stubs yield direct cost savings. For instance, Partners HealthCare saved $54,121.59 between 2002 and 2009, and conserved 2,845,743 of unprinted pages,[11] thanks to their success in connecting HIT priorities with sustainability imperatives and physician values.

> Kaiser Permanente's electronic health record system eliminated more than 1,000 tons of paper records and 68 tons of x-ray film, and [...] has lowered gasoline consumption among patients who otherwise would have made trips to the doctor by at least three million gallons per year.[12]

Leadership dialogue

1   What strategies could healthcare leaders use to help physicians appreciate that their patients and families evaluate them, in part, on how they relate to environmental stewardship?
2   What conditions relating to sustainability might a healthcare leader include in contracts to ensure physicians understand and practice the hospital's sustainability values?
3   How can healthcare leaders help physicians recognize that positive sustainability performance by the hospital supports their professional imperative to improve health?
4   How can leaders use energy-saving innovations involving information technology to engage physicians not only in optimizing clinical care processes and convenience, but also in achieving sustainability goals?

SUSTAINABILITY IN ACTION

Brenna Davis is former Director of Sustainability at Virginia Mason Medical Center, one of the most sustainable health systems in the US. She is a sustainability expert, thought leader and writer. Under her leadership, Virginia Mason Medical Center built its program into a globally recognized leader in healthcare sustainability. Brenna began her career in the energy industry, and has nearly two decades of sustainability experience. She is Chair and a co-founder of Washington Business for Climate Action, and has spoken on climate change across the state, at the White House and at COP21 in Paris. Brenna is author of *Regeneration: How the Resilient Heart, Creative Mind, and Compassionate Soul Light the Path to Global Sustainability* (forthcoming).

What inspired Virginia Mason Medical Center to become a leader for healthcare sustainability?

By supporting sustainability, we are supporting our communities. We believe in protecting patients of future generations from climate change. We want to move the healthcare industry in this direction, both nationally and internationally.

You have been named "One of the Greenest Hospitals in the US." How do you encourage other hospitals across the country to adopt green business practices? What efforts to encourage adoption have worked, and which haven't?

A few strategies have worked. We founded the Pacific Northwest Sustainability Leaders Roundtable, an information exchange program for industry. Many healthcare organizations in our region signed up to that group to share best practices and work on projects together. Healthcare Without Harm is our work on a national level. We also spoke at the Healthcare Climate Conference in Paris about how to encourage businesses to adopt green practices. I co-founded Washington Business for Climate Action, which comprises 260 businesses calling for action on climate change in Washington state.

Resilience in business and understanding health risks with climate change has resonated with people. Research has been done at Yale, George Mason University, the University of Michigan, etc., which showed that healthcare professionals are the most trusted public voices. It is so important for us to be change-makers and spokespeople around climate change.

What was unique about your position that allowed Virginia Mason Medical Center to be the first company in the industry to be outspoken on climate change?

I didn't know we were the first. Seattle has a very deep commitment to sustainability, with our composting, recycling, carbon neutral energy, etc. In Seattle, we know that climate change is a huge threat. Other Seattle businesses, like REI and Starbucks, were speaking up about this topic, and we were able to join the conversation.

We are in the healthcare industry, which gives us the moral obligation to talk about climate change. We were inspired to align with other businesses in Seattle around this mission. Climate change is the biggest human-made threat in our history. We also adapted the Toyota Production System to healthcare, and use it as our management method. This system allows us to eliminate wasteful processes from our patients' experiences and increase quality and safety.

What took the company on the sustainability journey and what were some obstacles you had to overcome?

Historically, sustainability and climate change happened outside of hospital work. We have always been focused on patient safety and care. When I came into this job, I looked at how sustainability supported patient health. The connection became more obvious the more I dug into it. But, originally, the topics seemed separate. I had to learn that sustainability and patient care go hand-in-hand.

Here are some highlights of our journey:

- 1980s: Recycling begins at our facility.
- 1990s: Energy efficiency issues kick off our conservation efforts.
- 2000s: Composting begins on campus, the first in the region.
- 2010s: Recycling begins in the operating room, the first in the region. The sustainability program expands to include a formal Green Team and Sustainability Director.[13]

What tools or resources have you found helpful in your journey toward sustainable healthcare?

Valuable resources in the industry include *Sustainability for Healthcare Management: A Leadership Imperative;* Kathy Gerwig's book, *Greening Health Care: How Hospitals Can Heal the Planet;* Practice Greenhealth resources, Healthcare Without Harm resources and a yearly GAP assessment.

Good resources outside of the industry include the Virginia Mason Production System/Toyota method, a project where we look at other industries to see how they are solving similar problems. We call it having "outside eyes" to help us see what may come next. This strategy has been helpful in allowing us to quickly test and adopt sustainability projects. We have a culture of change that is embedded in everything we do. Because we have that foundation, we can implement sustainability projects rapidly and measure what works and what doesn't.

Any advice for latecomers to sustainability?

We are moving into a resource- and carbon-constrained world. The businesses that are conserving and preparing for these restraints are setting themselves up for success in the next decade. Sustainability is no longer a nice-to-have add-on. It's an essential part of business. Given that reality, the first thing to do is a sustainability assessment of your business. It's okay to start small – for instance, by determining how to use less water. In coal-focused areas around the country and world, conservation has an even greater impact on human health. It's important to not think about conservation through the lens of blame, shame or judgment, but to think about sustainability through the lens of hope, innovation and economic sustainability. It's important to view this culture of sustainability as a business imperative. Rather than cut down people and businesses for their less than desirable environmental practices, let's inspire and empower a culture of sustainability.

Other comments

Our Lean and Green work: All of the sustainability projects we implement use "Lean and Green" meaning process improvement and environmentally sustainable operations. We just did an event on office supply recycling, where our admins had to throw stuff away in the trash versus recycling, etc. We discovered that we were buying $1 million in office supplies each year and could reduce our consumption and waste.

Syllabus integration

Overview

Healthcare today depends on information technology, yet not much attention has been paid to sustainable HIT. There will continue to be profound implications for sustainable HIT systems as healthcare fully transitions from a paper-centric environment to an IT-centric environment. This lesson will introduce opportunities for environmental HIT practices based on case studies.

Objectives

1   To study examples of how benchmarking results lead to the implementation of sustainable HIT improvement programs.
2   To provide a format, structure and step-by-step process for a benchmarking study.
3   To introduce a useful total quality environmental management (TQEM) tool.

Topics

1   Infrastructure investment
2   Storage management
3   Server virtualization
4   Print reduction and elimination

Homework

BSR (2016). *Measuring Environmental Benefits of Health Information Technology.* Available at www.bsr.org/our-insights/case-study-view/center-for-technology-and-sustainability-health-it-kaiser-permanente.

Brown, B. and Thiry, M. (2010). *Sustainability "How-to Guide" Series.* Houston, TX: IFMA Foundation.

Greening through IT: Information Technology for Environmental Sustainability. Available at www.ncbi.nlm.nih.gov/pmc/articles/PMC3040630/.

Additional resources

EPEAT® the definitive global registry for greener electronics. Available at www. epeat.net/.

Nimpuno, N., McPherson, A. and Sadique, T. (2009). *Greening Consumer Electronics – Moving Away from Bromine and Chlorine.* Göteborg: International Chemical Secretariat, and Spring Brook, NY: CPA (North America). Available at www. issuelab.org/resource/greening_consumer_electronics_moving_away_from_bromine_and_chlorine.

Practice Greenhealth (2009). Greening operations series: change is good! Using change management techniques to improve environmental performance – with a case study on paper prevention. Webinar, October 16. Available at www. practicegreenhealth.org.

Notes

1 Gidlason, D.S. (n.d.). Public health vs. the doctor's office. Nutramed. Available at www. nutramed.com/medicalcare/public_health.htm.
2 Epstein, R. (1995). Communication between primary care physicians and consultants. *Archives of Family Medicine* 4(5): 403–9. Available at www.ncbi.nlm.nih.gov/pubmed/7742962.
3 Pho, K. (2011). Primary care doctors and specialists need to better communicate. KevinMD.com. Available at www.kevinmd.com/blog/2011/01/primary-care-doctors-specialists-communicate.html.
4 Beeson, S. (2006). *Practicing Excellence: Five Reasons Medical Groups and Hospitals Striving for Culture Change Must Get Physicians on Board.* Gulf Breeze: Five Starter Publishing.
5 Collins, A. (2011). Food, health and the environment: the role of the clinician. FoodMed. org. Available at www.foodmed.org/2011/presentations/E2-Collins.pdf.
6 Caper, P. (2009). Health care should be driven by mission, not money. Physicians for a National Health Program, December 7. Available at www.pnhp.org/news/2009/december/health-care-should-be-driven-by-mission-not-money.

7 Turley, M., Porter, C., Garrido, T., Gerwig, K., Young, S., Radler, L. and Shaber, R. (2011). Use of electronic health records can improve the health care industry's environmental footprint. *Health Affairs* 30(5): 938–46. Available at http://content.healthaffairs.org/content/30/5/938.abstract.
8 Ibid.
9 Terry, K. (2009). Hospital EHRs don't make paper go away. *CBS News*, July 29. Available at www.cbsnews.com/8301-505123_162-43840588/hospital-ehrs-dont-make-paper-go-away/.
10 Ibid.
11 Partners HealthCare System, Massachusetts, 2009 (interview).
12 Turley et al. (2011).
13 Fast Facts Virginia Mason – Sustainability (2016). Available at www.virginiamason.org/workfiles/enviromason/enviromason_fastfacts.pdf.

4

WASTE NOT, WANT NOT

Process improvement + waste management

FIGURE 4.1 Waste, waste, waste and more waste

Executive summary

Hospitals are places of healing and also producers of substances known to harm the public's health. When healthcare leaders speak of waste, they are often referring to inefficiencies in time or resources as opposed to solid waste. Applying Lean methods to process improvement is becoming increasingly popular in healthcare organizations. Parallel efforts within hospitals to reduce tangible waste are emerging, such as solid waste and sanitary sewer effluent.

Healthcare leaders have the potential to help employees and patients improve individual health through clinical processes while improving community health through waste segregation and alternative off-site waste management services. Wasting less intrinsically means saving more – a combination of waste reduction and prevention strategies is saving millions of dollars annually at healthcare organizations across the United States.[1] This chapter explains why process improvement, when combined with waste management practices, can improve efficiencies and reduce operating costs.

A told story

Georgia Trohv, a nurse in surgical services at Memorial Hospital, is passionate about the environment and has recently completed Lean process improvement training. She discovers that regular trash is being thrown into the regulated medical waste bin because the bin is large and convenient for nursing staff. Georgia works with Environmental Services and Materials Management to obtain smaller regulated medical waste containers that are appropriate for the amount of regulated medical waste generated daily. Georgia also works with her Unit Director, Franco Popellero, to conveniently locate municipal solid waste bins (regular trash) and to place regulated medical waste bins where they are easy to access as a second stop. These changes result in proper waste sorting through right-sized containers and strategic receptacle placement. Ultimately, the changes will lead to less costly waste management practices. When Georgia communicates with Franco and her nursing colleagues in other units, they too are excited to "do no harm" from both patient and environmental perspectives. This success story eventually gets to Fred, the CEO of Memorial Hospital, who meets with Felipe, the VP of Materials Management, to better understand the process improvement strategies in use.

Change the system: linking Lean

The core concept of reducing waste is fundamental to both sustainability thinking and Lean thinking. Similar to waste management, Lean is a management technique whereby work processes are redesigned to increase efficiencies and reduce waste. According to the Lean Enterprise Institute, "Lean means creating more value for customers with fewer resources."[2] Linking the Lean approach with organizational values, such as sustainability, is an important step. Lean focuses on efficiencies along the entire process, from concept to implementation to understanding customer needs. Lean works horizontally through an organization as well as along process verticals.[3] Like Lean, sustainability can improve a healthcare organization's competitive outcomes by reducing waste and operating costs.[4]

THE HISTORY OF LEAN PROCESS IMPROVEMENT

Process improvement has a lineage that stems back to the concept of interchangeable parts conceived by Eli Whitney. This milestone was followed by the work of industrial engineers in the 1890s. Henry Ford contributed further to process improvement by introducing manufacturing strategy. Society's current understanding of Lean is largely rooted in the Toyota "just in time" process.[5] As Toyota incorporated the Ford production model into its own production process, Toyota discovered opportunities for improvement. The work of Taichii Ohno and Shigeo Shingo recognized the value of human capital beyond physical labor in terms of quality improvement, concepts captured by James Womack in his highly regarded book on the subject, *The Machine That Changed The World.*[6] While Lean has a history in the manufacturing industry, the fundamentals of Lean are built on the premise of improving processes and quality. Lean principles have been applied to the healthcare setting in an effort to improve both clinical and non-clinical work process outcomes and the patient experience.[7]

What does waste mean in terms of process improvement?

As the figures in Chapter 2 illustrate, healthcare sustainability as a movement has grown significantly since the inception of Practice Greenhealth's Environmental Excellence Awards in 2002. Despite the positive progress, many healthcare organizations have yet to fully embrace sustainable practices, evidenced by the 7,000 tons of waste produced daily by the healthcare industry.[8] Waste reduction strategies often provide a starting point for health organizations newly energized around sustainability, because waste management can be highly visible (e.g., recycling) and there is a clear business case for sustainable waste management.

WHAT A WASTE!

The second largest expense on a hospital's balance sheet (following labor) is supply chain costs. As reported in 2009, the average hospital provider spends more than $72 million a year on supply chain functions, nearly one-third of its annual operating budget. The majority of the materials procured by a hospital ultimately become waste, resulting in $10 billion annually in disposal costs across the healthcare industry. By instituting waste management programs along with changes in material consumption patterns, a healthcare facility can reduce the amount of waste it generates and thus its waste handling and disposal costs as well as its environmental impact.

(www.sustainabilityroadmap.org/topics/waste.shtml)

Improving the sustainability of a healthcare organization relies on process improvement. "Linking environmental goals to business values is an important step towards promoting sustainability."[9] The greatest potential for improvement in any process is achieved when the boundaries of the process are expanded upstream and downstream so that the "upstream steps" and the "downstream impact" are studied across the entire continuum.[10] Improving processes across boundaries is challenging, as healthcare is often siloed along professional, departmental or facility lines. Healthcare culture is generally focused on making improvements within departments or service lines, as opposed to across the system, where the greatest opportunity for improvement lies.[11] The remainder of this chapter discusses process improvement in terms of managing tangible waste, taking the topic a step further to examine financial implications and de facto alignment of process improvement for sustainable waste stewardship.

SUSTAINABILITY IN ACTION

Iqbal Mian, MSSM, is the Member Engagement Manager for Practice Green-health. He manages a portfolio of healthcare members, and develops content and resources on energy and water conservation, clinical quality and various other topics. Mian received his Master of Science in Sustainable Management from the University of Wisconsin, Oshkosh. Since 2010, Iqbal has led sustain-ability initiatives, first at Affinity Health System, then at Ministry Health Care, and most recently at Ascension Health. He has developed policy, toolkits and sustainability strategies, and was in charge of implementation in multiple initi-atives, including energy conservation across the 15 hospitals and 46 clinics within Ministry Health Care.

How does Lean help develop Practice Greenhealth's sustainability and process improvement projects?

A lot of folks question Lean process improvement as a concept and methodol-ogy, and for good reason: It can be a confusing world with tons of terminology and data. To break it down, Lean is about incremental steps that remove waste in a system to create a better future state. The ultimate beneficiary of Lean is the customer, or the patient when applied to healthcare. At Practice Green-health, we have ten key topic areas that focus on improving the delivery and operations of hospitals in their efforts to be more sustainable. These topic areas are:

1　Leadership (structure and goals)
2　Waste (hazardous, solid, etc.)
3　Chemicals
4　Greening the OR
5　Food
6　Environmental preferable purchasing (EPP)
7　Energy
8　Water
9　Climate
10　Green buildings.

As we look at these categories through the lens of improvement, we use Lean principles to drive positive change that impact the patients and communities which hospitals serve.

　　For example, in the Greening the OR initiative by Practice Greenhealth, a baseline checklist is made available that includes 50+ items to help hospitals identify gaps in environmental performance (i.e., Are medical plastics recycled? Has the facility eliminated the anesthetic gas of desflurane, which is a powerful CO_2 contributor, or installed LED lighting?, etc.). The baseline checklist aids hospital OR staff to assess opportunities for improvements in a standardized way. The Greening the OR Initiative is a small example of how we guide members of Practice Greenhealth to remove the waste (both physical and process) to create healthier environments for patients and employees.

Perhaps one of best-known tools within Lean and the Toyota Production System is "A3 Problem Solving." The idea is to conduct the scientific method on an 11 × 17 A3-size sheet of paper; hence its name. On the left-hand side of the paper, state the problem or issue, current condition (how things are occurring in the system) in which the problem exists, and then ask the five Whys to perform a root-cause analysis. Once participants understand the problem, move to the right side of the paper to create the ideal target condition and solution (the new steps and countermeasures to resolve the issue) – all within that one sheet of paper. Using A3 problem solving is a simple, effective way to document sustainability projects while preventing over-complication. We borrow from Albert Einstein who was known for spending the majority of his time exploring the problem and only then implementing the solution. All too often, the tendency of problem solvers is to jump to a quick solution without fully comprehending the challenge. A3 problem solving is a powerful tool for anyone on the frontline of operations, as it helps identify potential barriers and reveal bottleneck(s).

As a general rule of thumb, leadership must support and empower employees on the frontlines doing the work that is closest to the patient or customer. Leaders who empower employees and advocate problem solving using Lean tools can dramatically transform the way teams and departments operate to create immediate value. Moreover, the utilization of Lean thinking helps transform culture and improve employee satisfaction over time.

What were the outcomes of the projects at Practice Greenhealth around sustaiinable, eco-friendly practices in healthcare?

Sustainability is a world of complexity. The ability to break down problems into manageable pieces through Lean methodologies has proven to be effective. To further help the industry in this space, PGH is focused on standardizing work. Each hospital has its own way of doing things. Creating standard work allows PGH to capture best practices and engage leadership to implement best practices across a facility, industry, region or nationwide. At Practice Greenhealth, we have been able to create tools and templates to standardize work and minimize variation while allowing room for the creativity that brings meaningful results.

For measures like creating a Green Team in a hospital, we provide a standard set of work (guidelines/framework), documented in a template with tables and columns. These columns include the following:

1   What: the high-level steps needed
2   How: the details to execute each step

3 Why: the reasons behind steps to help illustrate need or importance
4 Who: the position or person(s) involved in the step
5 Visuals and/or links: to help "see" the steps, for perspective.

This table of standard work instructions allows individuals, teams, departments and hospitals to scale efficiency on various sustainability projects and initiatives. Every year, PGH hears from member hospitals about sustainability progress within the ten topic areas outlined at the outset of this "Sustainability in action" interview. The outcomes of their work range in innovation and results. Across the board, hospitals reduce their overall carbon footprint and meaningfully engage employees in sustainability initiatives. Example sustainability initiatives include drawing up strategic plans, creating composting programs, providing alternative methods of transportation for employees (electric vehicles, bus to work programs, etc.) and challenging traditional design elements by creating spaces with more access to daylight and vegetative roofs or gardens. To learn more, check out the annual benchmark report from Practice Greenhealth.

Who were the stakeholders?

Simple answer: sustainability is about the patients and the community. Here at Practice Greenhealth, employees from the top down passionately come together every day to work on the ten topic areas and their respective projects and initiatives because they know that the major stakeholders who are impacted are those who are most vulnerable: the patients. When hospitals reduce energy use which in turn reduces emissions, it is understood that pollutants from burning fossil fuel sources are no longer contributing to respiratory issues. When building materials in a new construction space are manufactured without harmful chemicals it is celebrated that the health of employees, patients and the environment won't be negatively affected.

The key lesson to take away is that being conscious of Lean methodologies and principles in healthcare supports positive outcomes in sustainability (people, planet, prosperity). It's not rocket science, but it requires us to ask the right questions and be bold to challenge the status quo. Change isn't easy, but we know the results when we hear "That's the way we've always done it."

What does waste cost?

According to the Joint Commission Environment of Care Standards, "Hospitals lose money due to regulated medical waste management costs. Simple management changes recoup 10% to 50% of costs (and reduce pollution)."[12] There are a variety of reasons to reconsider waste management practices, including increased federal regulation, public pressures and increased costs. A major reason to think differently about waste management is that many current models of waste segregation result in wasted money:

Throwing away non-contaminated waste into the regulated medical waste stream (red bags) may increase regulated medical waste volumes by as much as 50 per cent! Encouraging proper waste segregation and selecting products that don't wind up in the trash can safely reduce regulated medical waste volumes and save 40 to 70 per cent on waste disposal.[13]

Waste management can be a community health benefit because it reduces the spread of toxic chemicals. But what is the cost of not managing waste sustainably? Historically, no one gave much thought to the opportunity cost of ignoring waste management, but in today's economic environment, healthcare organizations are looking for ways to cut costs and make care more affordable. A healthcare organization that reduces its physical waste stream reduces its waste disposal costs.

Table 4.1 reflects waste costs by region in the US as of 2010 (see also Figure 4.2).

FRED: You're telling me that we'll reduce costs if we sort our waste properly and manage how we dispose of or recycle the different waste streams?

FELIPE: That's right. You may be surprised to learn that our competitor saved $1.2 million last year by reprocessing medical devices.

FRED: What are a few examples of items we can get started on with regard to reprocessing?

FELIPE: One of the most financially lucrative items is a harmonic scalpel. Other commonly reprocessed items are orthopedic burrs and bits and, in general surgery, laparoscopic surgical devices, pneumatic cuffs and ultrasonic scalpels. Some of these products are solid metal devices that are otherwise thrown away after just one use.

FRED: You mean we'd use a medical instrument for a procedure with one patient and then reuse the same tool with another patient? Is that safe?

TABLE 4.1 Waste costs by US region[14]

	Average costs by region ($)						
	California	Northeast	Midwest	Southeast	Southwest	Mid-Atlantic	Northwest
Solid waste per ton	158	122	90	53	109	136	150
Recycled per ton	51	41	88	82	29 (revenue)	83	25
Regulated medical waste per ton	1,111	1,765	821	552	461	729	1,618
Hazardous waste per pound	4.99	4.66	1.91	0.27	1.76	3.32	6.61
Number of respondents	12	13	21	2	3	11	11

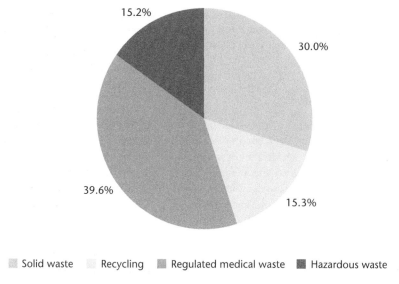

15.2%

30.0%

39.6%

15.3%

☐ Solid waste   ☐ Recycling   ☐ Regulated medical waste   ■ Hazardous waste

FIGURE 4.2 Average percentage of total cost

FELIPE: When a medical device is reprocessed, it means it gets sterilized before the device is reused. When it gets reprocessed, the device is completely dismantled into its individual components. Each of those components is then sterilized and tested. Lastly, every complete device is function tested, which is a higher rate of testing than brand-new devices receive.

FRED: You mean to tell me that testing for reprocessed devices is actually more stringent than testing for new devices?

FELIPE: That's right. That's exactly what I mean to tell you, Fred. There's a natural alignment between the cost strategy and the sustainability strategy. If you cared not a whit about sustainability, and you were solely focused on driving down the cost profile of operations, then you should be concerned about waste management. The cost of waste sorting for our entire hospital comes to an average of a quarter of a million dollars each year. Granted, waste management is just a small percentage of the total operating budget, and that percentage of our total costs hasn't changed much over the years. But, there is no reason to throw cash in the trash. I suspect there's a lot beyond the cost implications that you probably care about with regard to waste management.

FRED: The conversation about waste goes something like this: high waste means low quality and high cost. Every leader I know wants high quality and low cost. So we want to drive out waste. Now I tend to be a cynic, and I think that reducing waste is a means of reducing cost from a sustainability point of view. But I understand that for some of our employees, sustainability is an end in and of itself, rather than a means to an end. The good news is that I imagine that many of our employees recycle at home, so recycling at work would be natural for them.

FELIPE: [*thinking to himself*] I wonder if there will come a time when sustainability as a program isn't a cost-neutral cost center [...] I wonder, if sustainability becomes a cost center that isn't cost neutral or isn't saving money, would it still make sense to have it?

Opportunity costs

There simply isn't enough human or financial capital to invest in every leadership opportunity. When leaders establish priorities, they effectively deny a focus on other opportunities. When healthcare organizations choose not to manage waste, it is important to examine the consequences. Consider the implications of no waste management for construction and demolition debris (CDD) and its associated cost. During major capital improvements, hospitals face the decision of recycling CDD or letting it go to solid waste landfills. CDD is typically the largest volume of waste that goes into landfills.[15] When healthcare organizations divert waste, they not only have the opportunity to recycle resources, but many hospitals make donations to less developed countries. This practice benefits the recipients while keeping waste out of landfills.[16]

Commonly donated items from CDD include the following:[17]

* Clinical tools
* Medical equipment
* Furniture
* Computers
* Linen
* Books.

Given the relative ease with which these items can be donated, leaders should be asking themselves: what is the opportunity cost of not implementing waste management, not just for CDD, but across all operations?

FRED: Sustainability accomplishments aren't necessarily going to serve cost reduction goals, right? I mean, there's a chance that doing the right thing for the environment could actually cost us more, I presume? Let's face it, many sustainability projects require some upfront resources – whether it's time, attention or money. So what's the rationale for sustainability goals?

FELIPE: Well Fred, we're used to thinking about a single bottom line: cost. But sustainability is about a triple bottom line, as you know: environmental, social and fiscal responsibility. In today's competitive healthcare marketplace, we often make decisions based on cost and safety. The trouble is that when we focus solely on cost, we often miss opportunities to achieve a triple bottom line. We overlook potential solutions that incorporate environmentally or socially responsible practices. Making decisions based on cost alone may be the most fiscally responsible thing to do from a narrow perspective. However, that mindset neglects our broader mission to fulfill social and environmental responsibilities. Social and environmental outcomes have consequences and costs for all of us as members of the global community.

FRED: We rarely drive decision making by focusing solely on cost, Felipe, despite how it may appear.

FELIPE: Of course, Fred, I understand that. Let me make clear that I don't think it's okay to neglect cost efficiencies to the benefit of socially and environmentally responsible operations. Sustainability is about optimizing and balancing the intersection of all three bottom lines – social, environmental and fiscal. Focusing on one bottom line at the expense of the others won't help us much in the long run.

FRED: I see. I gather that when the benefits of two of the bottom lines outweigh the third bottom line, it could still be a high-value proposition from a sustainability perspective.

FELIPE: Potentially, Fred. That's right.

The human impact of waste management

Waste management has gained a level of sophistication in the past decade. Modern food service would be a good metaphor, which requires not only managing food production but also managing safety, nutritional value, customer service and cost. Twenty-five years ago, hospital food was a utilitarian exchange with less focus on taste.[18] If employees or hospital patients were actively dissatisfied, that was a problem, but expectations in the hospital setting for food management were low. Today, patients and employees expect the food in healthcare to be part of what drives a positive patient experience.[19] Similarly, waste management is experiencing a shift in how it is perceived. Patients and visitors expect to see recycling bins. A leader can do everything else right in terms of sustainability, but if he or she does not have recycling options, there is a negative image exposure in the public's eye. Recycling is the most visible and emotionally connecting element of environmental sustainability because individuals are engaged at a personal level.

There is no clear definition of where sustainable waste management accountability lives within the modern healthcare organization, though Environmental Services most often owns it. Outsourcing trends across hospital departments may result in gaps in management oversight.[20] Waste management is no exception, particularly since Environmental Services is often outsourced.[21] Sustainability is an essential component of the way in which every healthcare organization operates, and cannot be optional for leadership. Hospitals are responsible, both legally and socially, for their waste.[22]

FRED: I can't say I've ever given much thought to waste sorting, but we spend an average of a quarter of a million dollars on waste management each year[23,24] (Inova, 2013). If we could cut a fraction of that cost through waste sorting, that would be good. If we bought new supplies and then reprocessed them, we could save another chunk of change. I can't remember a time when we turned to waste management to help defray overall operating costs. That's a big change, and a big plus for sustainability.

FELIPE: Oftentimes, the cost of a reprocessed device is half the cost to purchase a new device. Our hospital model for the longest time has been that as prices went up, volumes went up, and expenses went up. So we weren't exactly forced to find cost savings. Meanwhile our friends in the airline industry, banking and telecommunications had to find 5 or 10 percent each year to keep consumer pricing competitive.

This is clearly an unsustainable model; hence the current focus on cost reduction in healthcare.

There are several waste streams in most hospitals, as Figure 4.3 illustrates (see also Figure 4.4). There are best practice standards for each type of waste stream (Table 4.2).[25]

Waste measures include the following:[26]

- average pounds of total waste per adjusted patient day (APD);
- average pounds of total waste per licensed bed per day;
- average pounds of total waste per staffed bed per day.

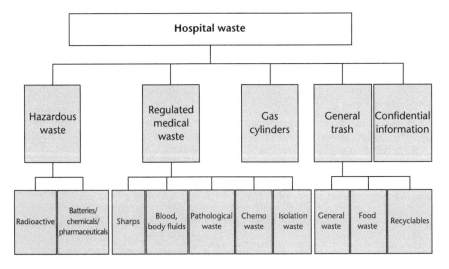

FIGURE 4.3 Hospital waste streams

TABLE 4.2 Waste streams and best management practices

Waste stream	Best practice (percentage of total waste weight)
Solid or non-regulated medical waste	60–70
Recycling	30–50
Regulated medical waste	8–10
Hazardous waste	>1

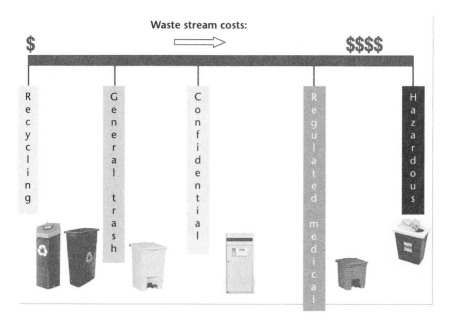

FIGURE 4.4 Relative waste stream costs

Getting paid for process improvement

The fiscal savings from waste management incentivize leaders to champion sustainable management practices. When leaders realize the cost savings of optimal waste disposal, the social and environmental goals that are equally important to the triple bottom line of waste management suddenly seem more achievable as well.

Most innovation is not about having ideas: it is about empowering people

They say that only a baby with a dirty diaper likes change. The truth is that no one likes change very much. For years, successful healthcare organizations have been rewarded for making incremental change.[27] This is especially problematic for organizations that are used to success, because they are likely the ones that need to change most if they are serious about engaging a new model of health.

How does a leader effectively engage employees and physicians in innovative waste management practices and get positive results? Most innovation and positive change does not come about because people have ideas, but rather when employees and leaders are empowered to act upon those ideas. Leaders must be explicit about empowering innovation around sustainability. The reality is that very little about waste management is truly innovative. But an organization's approach can become increasingly sustainable by making it easy for employees to experiment with and implement new approaches toward sustainable waste management.

Leaders must involve employees to weave sustainable waste management practices into daily operations. Employees have a responsibility to study and improve the processes in their department and teach colleagues sustainable waste disposal practices. Only after building connections with employees based on shared values (like sustainability) can changed practices begin to be hardwired into the organizational culture. At that point, sustainability becomes a purposeful element of every employee's job.

Sustainability accelerates strategy

There are a growing number of examples of healthcare leaders who have successfully hardwired sustainability into their organizations.[28] Several such organizations are highlighted in the practical applications within each chapter of this book. Among these leading organizations, some are known for sustainable food practices while others are known for sustainable facility design. All of the leading organizations share strengths in properly managing and segregating biohazardous from non-biohazardous waste materials. Red bag waste reduction is an area of focus for sustainability efforts that not only reaps financial savings, but is also the socially and environmentally responsible thing to do. Single-use device reprocessing, reusable sharps container programs and pharmaceutical waste management have become top priorities for leaders as an increasing number of hospitals implement programs to properly handle complex waste streams. Hospitals have in particular intensified their efforts to understand how to dispose of both hazardous and non-hazardous pharmaceutical waste sustainably.[29] A model pharmaceutical waste program is described below.[30]

1   Implement a pharmaceutical waste management program
2   Hire an outside vendor to help set up your program
3   Separate pharmaceutical waste at the point of generation
4   Send pharmaceutical waste back to pharmacy for proper segregation
5   Collect all pharmaceutical waste at the waste collection point and sort in a satellite accumulation area.
6   Treat all pharmaceuticals as hazardous waste.

FRED: Please have copies of the pharmaceutical waste management process for Monday morning's meeting so that we can facilitate an introductory conversation.
FELIPE: Sure thing.
FRED: Felipe?
FELIPE: Yes, Fred.
FRED: I suppose we ought to do the sustainable thing and send the document electronically instead of printing paper copies.
FELIPE: Will do, Fred.

Embedding the concept of waste

A key leadership role is to help colleagues understand alternative practices that are more sustainable than existing practices:

Given the potential to reduce costs while improving public image, waste is often the starting point for hospitals' sustainability efforts. Typically, hospitals start by getting a baseline picture of their waste and material streams and current systems in terms of waste volumes, costs, and management processes. If regulated medical waste (RMW) comprises more than 10 percent of total waste, red bag reduction is typically an early priority, due to its significant cost savings potential. Recycling is also an obvious opportunity for cost savings and environmental improvement.

(Practice Greenhealth, 2016)[31]

Many of the organizations that lead the healthcare industry in recycling practices make money from their recycling programs (the best case was over $181,000 annually). The other half of best practice hospitals for recycled content lost money (worst case: almost $88,000).[32] Regardless of the cost savings or expenditure, hospitals with developed waste management programs generate less waste than their counterparts at about a 30 percent rate (Figures 4.5 and 4.6).[33]

DIOXINS

Dioxins (waste by-products of industrial processes) are a community health concern because they can cause cancer, weaken the immune system, interfere with the endocrine system and reduce fertility. "The United States Environmental Protection Agency's 1995 dioxin emissions inventory estimated that medical waste incineration was the nation's third largest dioxins source, emitting 15 percent of all the dioxins on the national inventory."[34]

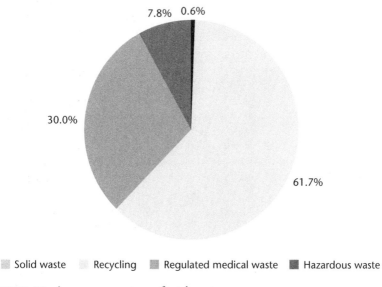

7.8% 0.6%

30.0%

61.7%

▨ Solid waste   ▨ Recycling   ▨ Regulated medical waste   ■ Hazardous waste

FIGURE 4.5 Average percentage of total waste

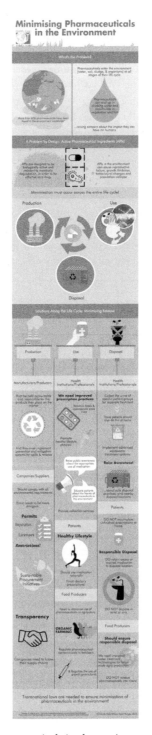

FIGURE 4.6 Minimizing pharmaceuticals in the environment

Low waste means high quality and low cost

In the United States, only 12.5 percent of electronic waste is sent to recyclers and the remainder is dumped or burned.[35] Even electronic waste (such as computers, monitors and medical devices) that is sent to recycling services may not be properly handled. According to e-Stewards, recyclers send an estimated 70 to 80 percent of their electronic waste to less developed countries, where it is burned in large piles for metals, exposing people and the environment to a host of toxins.[36] According to the United Nations, half of the world's population lacks access to safe waste disposal practices (UNNC, 2013). To avoid environmental, human health hazards and moral quandaries, ask for e-waste recyclers or ask group purchasing organizations for e-waste recyclers that have taken additional steps to be environmentally responsible.[37]

Electronics are not the only recyclable product for which leaders should find new options. There is also substantial opportunity in the area of reusable linens.[38] Other reusable items in healthcare facilities include the following:

- Surgical drapes
- Surgical gowns
- Incontinence products (underpads and briefs)
- Isolation gowns
- Scrubs
- Surgical packs (sterile and non-sterile)

KNOW BEFORE YOU THROW

Red bag reduction continues to be an area of focus for sustainability efforts. Hospitals that reap savings from these efforts report that they:[39]

- Engage in a regulated medical waste education and reduction program.
- Post waste segregation posters at red bag collection areas.
- Use a fluid management system in operating rooms.
- Use the fluid management system exclusively in orthopedic operating rooms.
- Use single-use device reprocessing.
- Implement a reusable sharps container program.

Figure 4.7 is an example of a waste segregation poster for a red bag collection area, available free online.[40]

FELIPE: You know, Fred, I went to the supermarket the other day, and when it came time to pay, the lady working the cash register asked if I'd like to make a dollar donation to a good cause. She asked the person ahead of me in line that same question. It struck me that the brilliance of this fundraising tactic was that it was hardwired in the checkout process. In other words, it was excellent product placement and logistically feasible.

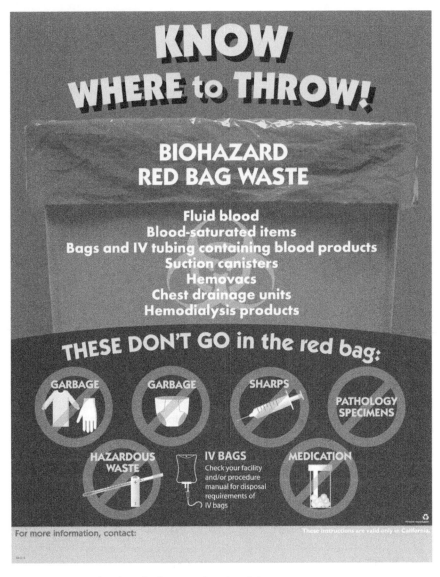

FIGURE 4.7 Regulated medical waste educational signage

FRED: You mean you could make a positive contribution to society while paying for groceries?

FELIPE: That's right, Fred. The supermarket made it easy for me. They made me feel like I was part of the solution. I think those are the two main reasons why I donated. I don't even remember what the cause was that my money went to.

FRED: So maybe if we put the recycling bin in easy-to-access, line-of-sight locations, our employees will be more apt to recycle. Do you think that might work? I'm convinced we'll have better results if we make it easy for our employees and visitors to think about managing waste sustainably. One of our biggest challenges will be getting the entire physician and employee population to build sustainability into the way we operate. They're so used to managing waste one way.

FELIPE: How about this? I'll walk the entire hospital and ask the stakeholders, our nurses, physicians and employees, what they throw where. We'll see if we can come up with a hardwired solution to sort our waste properly.

FRED: Sounds like a plan to me. I like the way you're engaging our colleagues to be part of designing the solution.

Practical application: providence's blue bag recycling program

The imperative

- Increasing recycling rates.
- Decreasing waste management costs.
- Circumventing restrictions on what local jurisdictions will accept from hospitals.
- Providing employment to underemployed social groups.

The initiative

Before Providence Health & Services began its focus on recycling, the system's hospitals were recycling small amounts of materials here and there, but lacked a comprehensive program. They hovered around a 15 percent recycling rate, but had aspirations of achieving a rate of over 50 percent. However, they could not find a vendor who would provide a unilateral, single-stream recycling service to their hospitals.

The person in charge of the system's sustainability initiatives took the issue to a local consortium of hospitals and industry vendors, through which they were able to develop an internal system for self-sorting the materials. They then partnered with a local organization that worked to provide jobs for developmentally disabled adults, using these individuals as helpers in sorting the recycled waste from its commingled containers. The hospitals packaged the separated materials themselves and sold them to local vendors, receiving a modest profit per bale.

After establishing this behind-the-scenes system, the focus turned to the front-line. In order to increase the amount of materials recycled, it was necessary to engage and educate the staff about Providence's recycling program. This process was achieved through a number of mechanisms:

- Presenting at large department meetings.
- Placing clear, visible signage on or immediately above containers.
- Providing peer-to-peer education and pressure from Green Team members.
- Putting together presentations highlighting specific problems when sorting issues occurred.

- Highlighting dollar savings from programs.
- Focusing on nursing staff, who do most of the sorting.
- Auditing, evaluating and correcting the placement of containers to make sorting easier.

The key individuals involved in the implementation of this project included property managers, the Director of Hospitality (who oversaw environmental services), Green Team members, quality practice councils, infection control, public relations and nursing staff.

The metrics

- Recycling rate
- Quantity of material recycled (lbs, tons, total bags)
- Dollars saved due to landfill diversion.

Lessons learned

- Share what colleagues are doing and the positive results they are seeing in order to garner administrative and clinical support.
- Engage people's competitive edge to challenge colleagues to reach further than they otherwise may in achieving sustainability goals.
- Highlight what will be "left on the table" if the organization does not pursue the sustainability program to spur engagement and productivity.
- Help staff understand what is and is not recyclable by sharing consistent information in as many different formats as possible. Particular attention must be paid to problem areas such as what to do with HIPAA (Health Insurance Portability and Accountability Act)-protected information and opened/unused supplies. This strategy is the most effective method to combat contamination issues. When implementing a commingled recycling program, education is key.
- Partner with local hospitals or meeting groups to identify vendors who accept commingled recycling in the healthcare setting. Every state in the US and virtually every country have these meeting groups for healthcare sustainability – they often have online forums for idea sharing and they often invite vendor partners to participate.

About the organization

Providence Health & Services is a not-for-profit Catholic healthcare ministry committed to providing for the needs of the communities it serves, especially for those who are poor and vulnerable. The health system spans five states – Alaska, California, Montana, Oregon and Washington – and includes 27 hospitals, 214 physician clinics, senior services and supportive housing, a health plan, a liberal arts university, a high school, and many other health and educational services.

Sources

Inova (2013). *2013 Sustainability Report* (Rep.). Fells Church, VA: Inova.

Mike Geller, Providence Portland Service Area Sustainability Coordinator, Providence Health & Services. Available at www2.providence.org/phs/Pages/default.aspx.

Practice Greenhealth (2016). *Sustainability Benchmark Report*. Available at https://practicegreenhealth.org/tools-resources/sustainability-benchmark-report-0 (accessed August 10, 2017).

United Nations News Centre (UNNC) (2013). Recycling, proper waste treatment can be veritable "gold mine" – UN environmental study. October 7. Available at www.un.org/apps/news/story.asp?NewsID=46211#.WX8L74TyuM8 (accessed August 10, 2017).

Leadership imperative

Why should a healthcare leader care about waste management? Beyond employee engagement, sustainability effectively translates to the executive level when it is integrated into the business goals of the organization. Use Lean as a vehicle for integrating sustainability to improve processes. Lean helps set and monitor targets and milestones that allow leadership to gauge sustainability progress. Lean helps make visible to leadership both the goal and the performance process. A healthcare organization may have the most environmentally progressive program, but the program will likely go unnoticed until employees and patients actively participate. Employees and patients alike relate to what is immediately identifiable.

Leadership dialogue

1   When combined with waste management practices, why and how does process improvement increase efficiencies and lower operational costs?
2   Why is it important to promote sustainability by linking sustainability goals, such as waste management, with business values?
3   Why does Lean thinking naturally align with waste management?
4   How can leaders effectively link the Lean approach with sustainability values for employees who do not recognize the natural alignment?
5   How can leaders improve waste management processes across professional, departmental and/or facility boundaries?

Syllabus integration

Overview

This lesson introduces an overview of process improvement as related to solid waste regulations, risk assessment methodologies, solid waste collection, disposal practices and recycling. Students will learn about waste minimization and pollution prevention, treatment of hazardous waste and remediation.

Objectives

1   To participate in a process improvement exercise.
2   To explore major sources and types of healthcare waste.
3   To identify opportunities to reduce waste.
4   To evaluate costs and benefits of proposed waste management practices.
5   To analyze gaps in current knowledge concerning waste management.
6   To discuss current legislation and regulation regarding waste issues.
7   To debate under what circumstances it is safe to reuse disposable items.

Topics

1   Types of waste
2   Waste subtopics
3   Halogenated anesthetic gas exposure
4   Pharmaceutical waste management
5   Reduce, reuse, recycle, prevent

Homework

Healthier Hospitals (2017). *Less Waste Challenge.* Available at http://healthier hospitals.org/hhi-challenges/less-waste.

World Health Organization (2017). *Health Care Waste Management.* Available at www.healthcare-waste.org/.

One of the most challenging efforts of healthcare sustainability is selecting a starting point. Reflecting upon classroom discussion and assigned reading content, identify where you would begin as a Sustainability Director located in a regional medical center, and why you chose that starting point. In addition, in your career journal, reflect upon how waste management practices impact human health and vary across the world.

Additional resources

Chen, I. (2010). In a world of throwaways making a dent in medical waste. *New York Times,* July 6. Available at www.nytimes.com/2010/07/06/health/06waste. html?pagewanted=all&_r=0 (accessed October 22, 2012).

World Health Organization (2014). *Safe Management of Waste from Health-care Activities.* Available at http://apps.who.int/iris/bitstream/10665/85349/1/9789241 548564_eng.pdf.

World Health Organization (2017). *Training Modules in Health-care Waste Management.* Available at www.who.int/water_sanitation_health/facilities/waste/training_ modules_waste_management/en/.

Notes

1 Brown, J. (2010). *Benchmarking Sustainability in Health Care Awards Benchmark Report.* Reston, VA: Practice Greenhealth. Available at https://practicegreenhealth.org/tools-resources/sustainability-benchmark-report-0.

2 Lean Enterprise Institute (2009). *What is Lean?* Cambridge, MA: Lean Enterprise Institute. Available at www.Lean.org/whatsLean.

3 Ibid.

4 Doman, J.L. (2007). Leveraging lean process improvement methodology to promote economic and environmental sustainability: obstacles and opportunities. MSc thesis, Department of Civil Engineering, Technology Environmental Management & Safety, Rochester Institute of Technology, Rochester, NY.

5 Lean Enterprise Institute (2009).

6 Anon. (n.d.). A brief history of Lean. *Strategos Lean Briefing.* Available at www.strategosinc. com/just_in_time.htm.

7 HealthCare Value Network (n.d.). HealthCare Value Leaders. Available at https:// createvalue.org/networks/healthcare-value-network/.

8 AHA (n.d.). Waste. American Hospital Association. Available at www.sustainability roadmap.org/topics/waste.shtml.

9 Doman (2007).

10 Lantz, P., Lichtenstein, R. and Pollack, H. (2007). Health policy approaches to population health: the limits of medicalization. *Health Affairs* 26(5): 1253–57. Available at http://content.healthaffairs.org/content/26/5/1253.full.

11 Lisha, Lo (2011). *Teamwork and Communication in Healthcare.* Edmonton: Canadian Patient Safety Institute. Available at www.patientsafetyinstitute.ca/en/toolsResources/ teamworkCommunication/Documents/Canadian%20Framework%20for%20Teamwork %20and%20Communications%20Lit%20Review.pdf.

12 Association of Bay Area Governments (2003). Why are hospitals rethinking regulated medical waste management? Available at www.abag.ca.gov/bayarea/dioxin/pilot_projs/ MW_Background.pdf.

13 Ibid.

14 Brown (2010).

15 Ibid.

16 Ibid.

17 Ibid.

18 Gordon, E. (2012). Fast food's slow exit from hospitals. *Kaiser Health News*, April 12. Available at http://capsules.kaiserhealthnews.org/index.php/2012/04/fast-foods-slow-exit-from-hospitals.

19 Watters, C., Sorensen, J., Fiala, A. and Wismer, W. (2003). Exploring patient satisfaction with foodservice through focus groups and meal rounds. *Journal of the American Dietetic Association* 103(10): 1347–49. Available at www.ncbi.nlm.nih.gov/pubmed/ 14520255.

20 Paris, L. (2011). Pew finds serious gaps in oversight of US drug safety. *Pew Health Group Safety*, July 12. Available at www.pewhealth.org/news-room/press-releases/pew-finds-serious-gaps-in-oversight-of-us-drug-safety-85899367931.

21 Anon. (2011). Outsourcing sees stimulus effect. *Modern Healthcare*, August 11. Available at www.modernhealthcare.com/article/20100920/MAGAZINE/100919948.

22 US EPA (2012) Medical waste frequent questions. US Environment Protection Agency. Available at www.epa.gov/rcra/medical-waste.

23 Association of Bay Area Governments (2003).

24 Department of Health and Children (2005). *Waste Management in Hospitals.* Government of Ireland. Available at http://audgen.gov.ie/documents/vfmreports/49_Waste_ Management_in_Hospitals.pdf.

25 Brown (2010).

26 Ibid.

27 Towers Watson and National Business Group on Health (2010). *Raising the Bar on Health Care*. Available at www.willistowerswatson.com/DownloadMedia.aspx?media=%7B4A 024110–2738–42EE-8F14–7EF06F4B839D%7D.

28 Practice Greenhealth (2012). Practice Greenhealth 2012 Environmental Excellence Awards. Available at http://practicegreenhealth.org/awards/.

29 Brown (2010).

30 Ibid.

31 Ibid.

32 Ibid.

33 Ibid.

34 Association of Bay Area Governments (2003).

35 Button, K. (2016). 20 staggering e-waste facts. February 8. Available at http://earth911. com/eco-tech/20-e-waste-facts/ (accessed August 10, 2017).

36 Ibid.

37 Brown (2010).

38 Ibid.

39 Ibid.

40 Cal Recycle (2006). Medical waste posters and stickers. Available at www.calrecycle. ca.gov/publications/Documents/BizWaste/44105012.pdf.

5

THE LONG AND WINDING ROAD (LEADS ME HOME)

Employee engagement + transportation

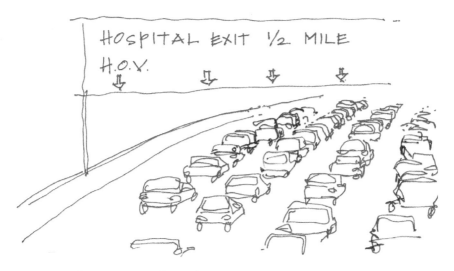

FIGURE 5.1 Commuting to a hospital

Executive summary

Engagement is critical to empowering employees and other key stakeholders in fulfilling the sustainability agenda. The outcome is predictably positive when leaders connect their organization's interests with the existing interests of key stakeholders, such as employees, physicians and the community. Those interests may be as diverse as the stakeholders themselves. Ultimately, sustainability is about personal health, community health and the health of the environment. This chapter explains why it is important to engage employees in the sustainability journey and how to do so, using transportation as the sustainability vehicle.

A told story

Gregor, Director of Safety and Security at Memorial Hospital, lives far from work and is plagued by rising gasoline prices. He arrives at the hospital stressed from sitting in traffic.

GREGOR: Gas prices are getting higher and my commute is not getting shorter. There has to be a better way to commute. Either that, or I need to find a job that is closer to home. [*The following day, Gregor sees a shuttle for a company drive by with one of his neighbors on board. After work, Gregor stops by his neighbor's house to ask how his company came to provide the shuttle service. Gregor learns that the local county government is committed to reducing traffic and provides funding to create shared ride programs. Gregor takes this knowledge to Denise, Chief Financial Officer, during a meeting about the construction of the new building.*]

GREGOR: I learned that the county is requiring us to have a plan to manage traffic congestion that is associated with construction. It's required as part of the building approval we obtained through the re-zoning process. I think part of the overall problem is that we aren't making it easy for employees to be sustainable about their commuting options. I for one am rather envious of my neighbor, who gets to ride to work each day in a shuttle. His company worked with the county to get funding for a shuttle from their community.

DENISE: That's interesting. Sounds like you've just found yourself a new project. Why don't you take the lead on researching what our options are for more sustainable transportation?

Employee engagement is a leadership priority

When a leader commits to sustainability, one of the immediate next steps is to build ownership among the leadership team and among as many employees as possible. Building ownership involves finding something that employees care about and embedding sustainability into that employee interest. Ownership starts to occur only when employees perceive that there is a "win" for them, or an alignment of interests between the hospital and themselves. Learning what employees value is a somewhat counter-intuitive place to start when operationalizing a sustainability agenda, but it is an important first step for the leader seeking alignment between employee values and organizational priorities.

At Inova Health System, for instance, 99.1 percent of the 545 employees randomly surveyed responded that "Environmental Sustainability ("Going Green") is important to them.

A different mindset

FRED: Do you think that a sustainability program would attract new employees? I could envision an applicant would interview at Memorial Hospital, hear about the sustainability program, and say, "Gee, that's different. I've been applying to jobs right and left and the sustainability program at Memorial Hospital stands out as forward-thinking. I think our values align."

FIONA O'MALLEY: That scenario is possible. Sustainability as a value may or may not be desirable from a recruiting standpoint. We'd have to ask HR.

FRED: The notion that sustainability is relatively new and therefore progressive is probably a very good thing. Sustainability attracts people who are looking to progress in their worldview, people who have relatively high tolerance for ambiguity and figuring things out. It's people who are progressive and change-friendly who largely define innovative organizations, because they create improvements that customers value faster than organizations that don't have change-friendly environments. The notion that it's new and therefore progressive is probably a very good thing. Sustainability attracts people who are looking to progress in their worldview, people who have relatively high tolerance for ambiguity and figuring things out. It's people who are progressive and change-friendly who largely define innovative organizations, because they create improvements that customers value faster than organizations that don't have change-friendly environments.

Embedding sustainability in leadership priorities

Any given hospital employs hundreds, if not thousands, of employees who move to and from the hospital each day. Visitors are also a major source of transportation impact. Less visible, but no less consequential for the community, is supply chain transportation, meaning transportation with the purpose of moving goods (i.e., medical products, food, waste, etc.) to and from the hospital. The leader who aligns organizational priorities with employee interests enlists valuable allies (e.g., reducing the cost or time required for the daily work commute). The ultimate goal, from a sustainability perspective, is to find tactics and approaches that improve the triple-bottom-line outcomes of lowering cost, improving the community's health and improving sustainability.[1]

Fred and Lori Prince of Human Resources engage in conversation.

FRED: Are we wrestling with transportation issues for transportation's sake, or are we're talking about transportation in terms of employee engagement? Our employees care deeply about how they get to work, especially in terms of time and cost. Frankly, it's a major source of dissatisfaction. I'm wondering if we can think about transportation in the context of sustainability.

LORI PRINCE: I don't think I've heard any of our leaders talk about transportation in the context of sustainability. I respond to employee complaints about insufficient parking, but that's about it.

FRED: One of our shared priorities is to strengthen employee engagement. What do you think about using sustainability issue A to help us achieve employee engagement goal B?

LORI PRINCE: I suppose it's worth a shot.

Besides employee engagement, transportation naturally overlaps with energy efficiency and cost reduction, clean air and related health effects, reduction of supply chain and waste disposal costs, and other environmental, fiscal and social issues. The key point for the purpose of this chapter is that there is a clear and explicit overlap between transportation strategy, employee engagement and sustainability.

My employees care about how they get to work

Recruitment has been a major issue historically and, given demographic trends, will likely continue to be an issue:

> Our nation already suffers from shortages of a range of health professionals, made worse by geographic maldistribution. [...] The shortages are expected to worsen as 78 million baby boomers begin to hit retirement age in 2011 and require more care for chronic illnesses.[2]

It is anticipated that there will be 1.09 million nursing vacancies by 2024 (Rosseter, 2017). Access to transportation is a key part of the recruitment equation in today's society. Public transit stops, highways and bike access influence the accessibility of the hospital to the workforce as well as patients.

Another trend impacting healthcare transportation in a significant way is the transition to home and ambulatory care.[3,4] Home health is one of the fastest growing sectors in the healthcare industry. Over seven million patients are served in home health settings each year. It is often the lowest-cost option for delivering healthcare services that would otherwise be provided in the hospital setting.[5,6] There is an anticipated growth of the home healthcare system of 70 percent by 2020. This is in comparison with 14 percent for the general US labor market (Ford, 2014). Primary care, in particular, is increasingly retail oriented and located outside of the hospital for optimal patient access.[7] This means that more services will be physically located in the community rather than at the hospital. Fewer people will be venturing to the hospital as more people transition to ambulatory care sites. One example of an expanding ambulatory program in the US is the Program of All-inclusive Care for

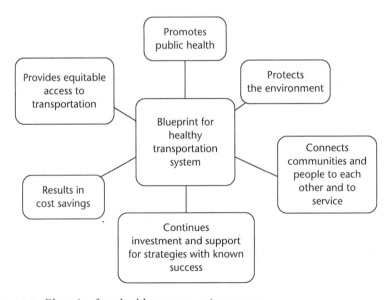

FIGURE 5.2   Blueprint for a healthy transportation system

the Elderly (PACE), a comprehensive service for enrollees that includes transportation of enrollees from their homes to the PACE site.[8] Alternatively, what are the transportation implications of moving large numbers (in small groups) of healthcare workers out to patients' homes? With an inadequate transportation system, delivering ambulatory and home healthcare will be affected and will likely lead to both dissatisfied employees and patients. As hospitals across the country embrace ambulatory and home healthcare models, the time is now ripe to incorporate sustainability into the transportation system.

LORI PRINCE: Have you given any thought as to where we should locate our ambulatory care sites and how smart growth plays into that decision?

FRED: I don't know enough about smart growth to have an informed opinion. But I have to say that, based on my limited knowledge, my reaction to smart growth has been negative. I mean, the basic concept of smart growth makes sense, you know, purposefully developing where people live and work in a contiguous space. But the way smart growth translates so often has sort of been against building roads or having any sort of investment in surface transportation. I never really understood how to apply smart growth to our interests.

LORI PRINCE: Well, Fred, I'm no expert on smart growth either, but I thought it was more about building density and public transportation. Proponents of smart growth are not against roads. Rather, they realize that unless we grow adequate density, we screw up land-use planning and public transportation won't make economic sense. And all that pollution will make all the pulmonary patients flock to the emergency department.

FRED: I'm out of my depth on the topic of smart growth, but I admit I never thought about it that way before.

To the degree that leaders engage employees in community issues such as transportation policy or air pollution rules, it is likely a relatively small yet growing piece of the total policy agenda. Support for sustainability, one could argue, is not much different from support for child safety seats, because they are both topics that everyone – patients, families, employees, physicians, etc. – is touched by. Improving transportation is directly related to improving health.

WHAT IS SMART GROWTH?

According to the US Environmental Protection Agency (EPA), smart growth is "development that serves the economy, the community, and the environment" (EPA, 2017). Smart growth is development that simultaneously achieves economic development and jobs, strong neighborhoods and healthy communities. Smart growth changes the terms of the development debate away from the traditional growth/no-growth debate to how and where new development may be accommodated.

Create connective opportunities

There are a wide range of higher order values that people care about, which emerge in finding a cure for cancer or pursuing a religious faith. For others it's engaging in sustainable actions or some combination of the aforementioned behaviors. Regardless of the topic of interest, connecting people with each other around higher order values is key to creating a culture of employee engagement. As sustainability becomes increasingly prevalent in healthcare organizations, employees will want to connect with each other and their leaders who embrace sustainability values. The challenge, then, is for leaders to connect their sustainability agenda with employees in a meaningful way.

What matters to employees? One objective of every nurse is to understand the patient's and family's value system. What will drive the family's behavior and the patient's compliance to manage health or follow the treatment plan? During the process of understanding the patient and family value system, it is likely that there could be a direct alignment between employees and patients who share a belief that sustainability matters. One can imagine a time when the nurse–patient conversation goes something like this:

> Hello. Please confirm for me your name and date of birth. We are going through this process for your safety. What would you like in terms of your diet? Was public transportation available to you? I am asking you these questions because we are trying to reduce our environmental footprint. These supplies you see are all recycled and meet the highest quality standard.

When employees articulate their motivation(s), whether it is safety or sustainability, they add value in the eyes of their patients and families. The patient thinks, "Hmm, my nurse cares about my safety, she cares about sustainability, she cares about me actually improving my health." Aligning values builds relationships and also builds trust. In this hypothetical situation, it is the core value of the caregiver, rather than the action taken, that provides the basis for alignment.

WHAT IS EMPLOYEE ENGAGEMENT?

Employee engagement is "a force that drives performance outcomes."[9] Successful organizations realize that, more often than not, their most valuable resource is their workforce. In addition, lost productivity has a cost, and a high one at that (Harter et al., 2010; see also Gallup, n.d.). It is estimated that $550 billion is lost from productivity due to unhappy workers. In light of cost alone, employee engagement is a leadership priority. Surveys show that the greatest success factor in getting employees engaged is through interaction with their management team (Murlis and Schubert, 2001).

TRANSPORTATION IS A HEALTH ISSUE

The way a community is designed and built has a lot to do with the future health of that community. The growth of suburban areas created lack of access to walking corridors and safe bicycle-riding routes, resulting in more auto-motive transportation. The additional burden and stress of sitting in traffic has an impact on health. Land-use planning and transportation planning are key components of the social determinants of health. Much of the land planning over the latter half of the twentieth century was based on the premise that people would own cars. The current reality is that many people cannot afford cars or choose not to use cars in urban settings and end up spending hours on round trips for groceries.[10] Lack of connectivity and access to healthy foods (i.e., food deserts) have added to the obesity epidemic which becomes a public health dilemma.[11] Well-planned communities are built around these access points.

In 2008, the government of the United Kingdom set out a strategy, "Healthy weight, healthy lives," that tied land planning with increased physical activity through walking and cycling.[12] Also published in 2008 was Public Health Guidance-PH8, which offered "guidance on the promotion and creation of physical environments that support increased levels of physical activity."[13] Of the seven evidence-based recommendations provided in this guidance docu-ment to the National Health Service (NHS), four focused on transportation access-related items.[14] The US Department of Transportation (DOT) has also recognized transportation as a public health issue. In 2010, DOT Under-secretary Roy Kienitz reinforced a statement about the impact which trans-portation policies have upon public health.[15]

How can transportation improve employee health?

Increasingly, healthcare employers seek to reduce their benefit costs and improve employee retention and productivity.[16,17] Often, employers attempt to achieve these goals by improving the health status of employees. Johnson & Johnson "leaders estimate that wellness programs have cumulatively saved the company $250 million on healthcare costs over the past decade; from 2002 to 2008, the return was $2.71 for every dollar spent."[18] Both Citybank and Bank of America estimated a saving of $4.5 for every dollar spent on employee wellness (NCSF, 2017). Component of many organizations' strategies is to embed opportunities for exercise into the work environment and work processes.[19] How can healthcare leadership connect the dots between transportation, exercise and employee health?

FRED: We have physicians and employees who bike to work every day. We know that the employees who exercise by virtue of transportation tend to be health-ier than their colleagues who do not exercise en route to work. Walking is such a simple form of transportation, it's often overlooked. Take internal transportation, for instance. Our employees could walk up the stairs instead of taking the elevator. I met recently with a Kaiser executive who told me that leaders at Kaiser are not permitted to use the elevators at their Oakland, California headquarters. Talk about a transportation policy!

LORI PRINCE: Do you know what our patients complain about most to our employees?

FRED: No, what? These questions get harder the deeper into this we go.

LORI PRINCE: It happens before they even get inside the hospital! Our patients and visitors complain most often about hospital parking or lack thereof. They complain that there isn't a covering from the garage to the hospital itself.

FRED: Maybe if we provided a covering, folks would be incentivized to walk. I see so many people take the shuttle option even though the garage is right across the street.

LORI PRINCE: I'm telling you, Fred, we probably get more complaints about how hard it is to get in and out of the hospital than anything else. Patients don't want to encourage their friends to visit them because the transportation and inaccessibility is so tough. That's one of the first messages our employees hear from our patients when they arrive at the hospital.

FRED: What do you think about labeling environmentally preferable parking spots in the garage for carpools and hybrid vehicles? I realize that doesn't solve the problem we were just talking about, but it does make our commitment to sustainable transportation visible to our community.

LORI PRINCE: I think that's a neat idea. What made you think of it?

FRED: Well, it was proposed originally as a way of encouraging people to buy hybrid vehicles. You know, there are a lot of people, especially younger employees, who are concerned about our posture *vis-à-vis* the environment. It's really all about Richard Florida's creative class principle. Have you ever read his book about the rise of the "creative class"? He says that the creative, bright, often highly educated people are attracted to organizations and communities that nurture the things that they find important.

LORI PRINCE: I know the type. Some of the things that this creative class finds important is individual liberty. They like customization. Employment policies, like having flexible work-at-home arrangements and open benefit structures that accommodate diverse lifestyles, are very important to them.

FRED: The point is that now I see how a small thing – like a parking spot – has symbolic value. Beyond that, it's an alignment of employee values with organizational sustainability objectives. One of the things that attracts the creative class is a clear commitment to sustainability and the environmental ethic.

Pursuing sustainable policies, transportation policies included, tends to be an employee satisfier. For many younger employees, engagement in sustainability practices is a minimum expectation of their employment.[20] Healthcare organizations that have bike racks, offer carpooling, shuttle options or other transportation alternatives will likely attract a meaningful number of employees.

The imperative

- Improve health by reducing transportation-related air quality issues.
- Reduce carbon dioxide emission by 3 tons per year through the integration of a sustainable transport system.
- Encourage participation of employees and patients through a heightened number of carpooling by 6,500 users in 2016.
- Support sustainable transportation.

The initiative

The World Health Organization recommends reducing greenhouse gas emissions to promote population health. According to statistics from the U.S. Environmental Protection Agency, 13 percent of greenhouse gas emission comes from vehicle exhaustion. Hospital carpooling services can support sustainable transport, optimizing the occupancy rate of vehicles and increasing the accessibility of rural populations to medical care. Carpooling passengers will help reduce high greenhouse gas emissions by cars through reduced vehicle use, hence leading to less air pollution.

Taichung Tzu Chi Hospital developed a mobile application to develop a sustainable transportation system. The hospital carpooling buses serve the community who would otherwise travel more than 20 kilometers (km) away from the hospital. Buses are linked with a customized mobile application. The application provides real-time information about the departure and arrival times of the buses for every stop. Passengers conveniently wait at the outpatient area once they arrive at the hospital. To encourage more hospital visitors to use sustainable transportation services, the hospital has developed brochures that are strategically placed within the hospital premises. To maintain and enhance the quality of transportation services, the hospital will need to spend around NT$1,400,000 (US$44,191). These measures enable Taichung Tzu Chi Hospital to maintain quality transportation services. The hospital invested approximately NT$100,000 (US$3,200) into the development of a carpooling services mobile application. The benefits reaped from the established system offset the costs.

The hospital established a team which consists of staff from the Medical Affairs Department and General Affairs Department. Hospital carpooling volunteers were trained and tasked to plan and adjust routes of the carpooling service. They are required to monitor the number of passengers who board at each stop. The data are used for tracking, personnel, and for analyzing the current system to identify possible improvements for the future.

The metrics

Carpooling services reduced the volume of fuel utilized to transport individuals and patients from the hospital to other strategic locations near their homes.

Since 2011, accumulated total savings of 1,688,412 liters for the 95 unleaded gasoline vehicles is NT$26,405,656 (US$833,512). The transportation system has served 277,534 passengers. Together, this system successfully offset 3,732 tons of carbon dioxide. Between December 2010 and July 2016, the hospital carpooling service transported 277,534 passengers. The transportation service provided enables the hospital to safely transport 250 persons per day to meet their medical needs. The Bus Transit System in particular services the patients and individuals living 20 km or further from the hospital. Ninety percent of the passengers were between the ages of 40 and 65, 36 percent had Oriental Medicine appointments, 22 percent had Neurology appointments and 16 percent had Gastroenterology appointments. The number of carpooling routes increased from nine to 12 in year 2013 with about 120 bus stops. The number of passengers served increased 1.7-fold in 2012. From 2011 to 2015, CO_2 emissions equivalent to 3,112 metric tons were conserved (4,893,082 kilowatt-hour).

Table 5.1 shows the volume of fuel conserved and the cost saved annually since its implementation (see also Table 5.2).

Lessons learned

Taichung Tzu Chi Hospital collaborated with community health centers through their routine health talks and workshops to promote the benefits of the Bus Transit System. During off-peak hours, when fewer passengers are using the transportation services, the frequency of pickups is adjusted. The local government learned about the need for public transportation to the hospital and subsequently reoriented

TABLE 5.1 Conserved fuel from the transportation system

Year	Volume of fuel in liters for 95 unleaded gasoline buses	Cost savings in new Taiwan dollars (NT$)
2011	45,825	714,622
2012	133,412	2,087,262
2013	234,070	3,658,018
2014	465,437	7,280,660
2015	529,364	8,278,302

TABLE 5.2 Carbon emission reduction from the transportation system

Year	Distance traveled (in km)	Total distance traveled (km)	Number of passengers	Carbon dioxide emission reduction (tons)
2011	27	458,255	16,844	101
2012	35	1,334,124	37,573	295
2013	36	2,340,705	65,027	517
2014	74	4,654,378	62,897	1,029
2015	88	5,293,640	60,155	1,170
2016 January to July	80	2,803,040	30,038	620
		16,884,142	2,77,534	3,732

the service route to avoid overlapping government resources with those of Tzu Chi Hospital.

Future plans entail developing routes according to the feedback collected from passengers and volunteers. The service expansions will be planned geographically (spaced a distance of every 10 km from the hospital). The hospital seeks to increase the engagement of the local community leaders, volunteers and other stakeholders in developing the Bus Transit System. These groups will join the hospital in identifying bus stops and ways to further improve the transportation service. With the use of technology to bolster human engagement, Tzu Chi Hospital is creating the most energy-efficient and carbon–emission saving means possible for transporting hospital visitors.

About the organization

Taichung Tzu Chi Hospital is located in Tanzi District of Taichung City (the central part of Taiwan). The hospital spans over 184,921 square meters and has 1,081 beds with more than 1,500 staff members. Tzu Chi Hospital provides holistic healthcare and preventive medicine services through community workshops, home visits and health counseling. Tzu Chi Hospital opted for local construction materials for its facility construction to protect the ecosystem and to prevent land erosion (links: http://taichung.tzuchi.com.tw/w/?id=1581).

Sources

This content has been adapted with permission from Global Green and Healthy Hospitals Initiative Case Studies, with permission.

FIGURE 5.3 Lining up for sustainability

WHAT ARE THE HEALTH IMPACTS OF TRAFFIC?

Transportation health impacts are both local and global in scope. On a local level transportation impacts air quality, which can have regional effects on airborne infections. On a global level, the potential exists for impacting the balance between the atmosphere, oceans and the sun, weather effects and effects on vegetation.[21]

Transportation has a wide-reaching impact upon public health. Driving is the most air-polluting activity for the average citizen.[22]

In fact, the US EPA has predicted the prevention of 230,000 early deaths by 2020 as a result of the Clean Air Act between 1990 and 2020.[23] An added benefit includes a projected $2 billion in savings from direct benefits of the program. Table 5.3 highlights some of the health areas of impacts from air pollution.

TABLE 5.3 Predicted reduction in health-related impacts from the Clean Air Act[24]

Health impact	Predicted number of cases	
	2010	2020
Adult mortality: particles	160,000	230,000
Infant mortality: particles	230	280
Mortality: ozone	4,300	7,100
Chronic bronchitis	54,000	75 000
Heart disease: acute myocardial infarction	130,000	200,000
Asthma exacerbation	1,700,000	2,400,000
Emergency room visits	86,000	120,000
Lost school days	3,200,000	5,400,000
Lost work days	13,000,000	17,000,000

Practical application: Cleveland Clinic's Alternative Transportation Program

The imperative

- Increase patient and employee satisfaction by eliminating traffic backups.
- Address environmental and health impacts of transportation.
- Reduce demand on automotive parking infrastructure.
- Minimize the commuting cost burden for employees.
- Pursue cost savings related to taxes on stormwater discharge and congestion in parking structures.

The initiative

Before Cleveland Clinic's transportation program evolved into the comprehensive program it is today, the program consisted of a variety of individual initiatives under the control of different groups within the organization: the wellness initiative promoting bike ridership, parking services promoting a carpool program and human resources promoting pre-tax public transit benefits.

Each initiative was independently overseen until the Chief Operating Officer and Chief Security Officer co-identified a large gap between growth projections and an available supply of parking spaces. They initiated a cross-functional work group to evaluate ways to reduce demand. Restructuring the organization's transportation program was a joint effort among protective services, the office for a healthy environment, the design and construction team, and the real estate team.

Cleveland Clinic's current transportation program consists of three main components:

- Providing commuting alternatives with the goal of increasing the value and benefit of Cleveland Clinic's public transit program to make it more attractive for employee participation. Examples of such programs included auditing and installing new bike racks, hosting transportation fairs to promote public transit options, and subsidizing public transit through a pre-tax public transit benefits program.
- Creating a sophisticated infrastructure by updating the technology used in garages, leveraging parking-related programs and building greener facilities. Examples of such initiatives included automatic vehicle identification (AVI) technology, pay-on-foot technology, creating flexible work arrangements to eliminate peak demand days, installing flexibility in parking rates to encourage part-time alternative commuting, and installing permeable pavers, bioswales and rain gardens to reduce the burden on stormwater discharge.
- Addressing fleet management issues to increase overall fleet efficiency. Examples of such efforts included developing purchasing standards to encourage the use of fuel-efficient or advanced technology vehicles, and developing anti-idling policies.

The metrics

Cleveland Clinic measures the success of its transportation program in a variety of ways, including:

- Overall fleet efficiency.
- Pounds of CO_2 emissions avoided due to wait and idle time reduced.
- Wait time coming out of buildings.

Results of commuter surveys:

- Quantity of people enrolled in carpool program and public transit program.
- Audits of bike racks to understand which are under- and overused.

Lessons learned

- Identifying an executive champion – such as the head of the transportation services team – is a crucial factor to achieve significant gains in overall fleet efficiency. Additional leadership involvement and commitment drives momentum and creates visibility.
- Shifting away from the "driving culture" is a tough sell in many places without a sophisticated public transit infrastructure. Direct communications coupled with appealing incentives help tackle this challenge. Communicate the *why* and *how* of using public transit, and then make it as easy as possible to participate by guiding employees through the process step by step.
- Cascade the message of sustainable transportation within the organization. Pursue inventive ways to convey the message. Messaging about the health impacts of transportation in a way that is tied to your organizational mission is a particularly effective way to make this topic resonate with employees.

About the organization

Cleveland Clinic is a nonprofit multispecialty academic medical center that integrates clinical and hospital care with research and education.

Sources

EPA (2017). About smart growth. February 23. Available at www.epa.gov/smart-growth/about-smart-growth (accessed August 10, 2017).

Ford, G.C. (2014). Home healthcare industry experiencing rapid growth. May 12. Available at www.thegazette.com/industry-experiencing-rapid-growth-20140511 (accessed August 10, 2017).

National Council on Strength and Fitness (n.d.). Corporate wellness programs return $6 for every $1 invested. Available at www.ncsf.org/enew/articles/articles-corporatewellnessprogramsreturn.aspx (accessed August 10, 2017).

Rosseter, R. (2017). Nursing shortage. May 18. Available at www.aacn.nche.edu/media-relations/fact-sheets/nursing-shortage (accessed August 10, 2017).

http://my.clevelandclinic.org/about-cleveland-clinic/default.aspx.

http://my.clevelandclinic.org/Documents/About/facts-figures-2010.pdf.

Chris Parkinson, Project Manager at Cleveland Clinic, Office for a Healthy Environment.

Leadership imperative

Sustainability was first introduced in this book by highlighting the triple bottom line. Every sustainability principle and most organizational priorities have a social, environmental and fiscal component. Transportation impacts employee engagement and patient satisfaction. Transportation also connects directly with supply chain management, an existing priority of most healthcare organizations. Supply chain management highlights the need to make explicit the invisible character of sustainability for those who consume supplies – healthcare as an industry is a leading consumer of supplies. Leaders tend to pay less attention to the transportation of materials, in part because the sustainability dimension is invisible to patients and employees. Today, buses powered by natural gas may feature signage displayed prominently on the vehicles to make the environmental benefits visible to passengers and people passing by. The same is true of hybrid vehicles that feature a prominently placed hybrid brand decal. Healthcare leaders can imitate these successful practices for supplies and equipment, thereby making otherwise invisible sustainability practices visible to stakeholders. The key is to make sustainable transportation processes visible while connecting sustainability to employees so that they are engaged in the organization's shared values.

FRED: Like I said before, the more visible the sustainability benefit, the better. For instance, you know that free-range coffee we have in the cafeteria? The point is that it has the sticker to indicate that the coffee is ethically sourced.
HR: Free-trade coffee.
FRED: Right. Free-trade. Give me some slack.

Leadership dialogue

1 Sustainability policies and practices within healthcare organizations are often compared with sustainability policies and practices of other industries. What metrics should healthcare organizations use to benchmark sustainability progress that are unique to healthcare versus borrowed from other industries?
2 Employee engagement is driven by a number of factors, including organizational culture and leadership expectations. Imagine that these two elements are not aligned as the leader implements a sustainability-oriented transportation policy. What is the subsequent impact upon employee engagement?
3 Who should influence sustainability-oriented transportation policies? The Board of Trustees? Management? The community? Elected officials? Employees? Why?
4 Are affinity groups (such as environmental interest groups) useful to engage employees in transportation and general sustainability issues? What are the pros and cons of encouraging affinity group development?
5 Why might a board be concerned about leaders becoming involved in community policy issues regarding the relationship of transportation and health? What are some possible implications for the hospital?

Syllabus integration

Overview

Students will assess and analyze issues related to sustainable transportation programs. The class will watch sustainable transportation video footage and will debate the ethics of sustainable transportation programs.

Objectives

1 Describe the environmental significance of transportation management.
2 Identify opportunities to efficiently manage transportation resources.
3 Explain why public/private partnerships are important to achieve sustainable transportation.
4 Assess why and how the community benefits from transportation management.

Topics

1 Factors that influence transportation operations
2 Risk categories of transportation management
3 Practical application promoting transportation management
4 Factors that influence the adaptability of sustainable transportation programs

Homework

Practice Greenhealth (2009). Greening operations series: worker commuting options and incentives. Webinar, December 18. Available at www.practice greenhealth.org.

Additional resources

NICE (2008). *Guidance on the Promotion and Creation of Physical Environments that Support Increased Levels of Physical Activity*. London: National Institute for Health and Clinical Excellence. Available at http://guidance.nice.org.uk/PH8.

Ross, A. (2010). *Planning, Transport, and Health Inequalities – A Recent History and Future Progress*. London: Local Government Association. Available at www. local.gov.uk/planning-transport-and-health-inequalities-recent-history-and-future-progress.

US Department of Transportation (2010). Transportation is a public health issue; DOT doing its part to keep kids moving. *Fast Lane*, July 16. Available at http://usdotblog.typepad.com/secretarysblog/2010/07/dot-doing-its-part-to-keep-kids-moving.html.

Notes

1 Elkington, J. (n.d.). From the triple bottom line to zero. Available at www.john elkington.com/archive/TBL-elkington-chapter.pdf.
2 Manos, D. (2010). Experts name 9 ways to fix healthcare workforce shortage. *Health Care Finance News*, January 15. Available at www.healthcarefinancenews.com/news/experts-name-9-ways-fix-healthcare-workforce-shortage.
3 McCain, M. (2011). *Ambulatory Care of the Future: Optimizing Health, Service, and Cost by Transforming the Care Delivery Model*. The Chartis Group. Available at www.chartis.com/files/pdfs/Ambulatory_Care_of_the_Future.pdf.
4 Gershon, R.R.M., Pogorzelska, M., Qureshi, K.A., Stone, P.W., Canton, A.N., Samar, S.M., Westra, L.J., Damsky, M.R. and Sherman, M. (2008). Home health care patients and safety hazards in the home: preliminary findings. Agency for Healthcare Research

and Quality. Available at www.ahrq.gov/downloads/pub/advances2/vol. 1/Advances-Gershon_88.pdf.

5 PACE (n.d.). State resources. National PACE Association. Available at www.npaonline. org/website/article.asp?id=203.

6 Gershon et al. (2008).

7 News Medical (2010). Sg2: demand for outpatient services to increase by nearly 22% over the next decade. *News Medical*, January 6. Available at www.news-medical.net/news/20100106/Sg2-Demand-for-outpatient-services-to-increase-by-nearly-2225-over-the-next-decade.aspx.

8 PACE (n.d.). Program of All-inclusive Care for the Elderly. Medicaid. Available at www.medicaid.gov/medicaid/ltss/pace/index.html.

9 Gallup (n.d.). Employee engagement. Available at www.gallup.com/consulting/52/Employee-Engagement.aspx.

10 US DOT (2010). Transportation is a public health issue; DOT doing its part to keep kids moving. United States Department of Transportation, July 16. Available at http://usdot blog.typepad.com/secretarysblog/2010/07/dot-doing-its-part-to-keep-kids-moving. html.

11 Policy Link (n.d.). The Food Trust. Available at http://thefoodtrust.org/food-access/publications.

12 Ross, A. (2010). Planning, transport, and health inequalities – a recent history and future progress. Local Government Improvement and Development. Available at www.idea.gov.uk/idk/core/page.do?pageId=23454074.

13 NICE (2008). *Physical Activity and the Environment.* London: National Institute for Health and Clinical Excellence. Available at www.nice.org.uk/guidance/ph8.

14 NICE (2008). *Guidance on the Promotion and Creation of Physical Environments that Support Increased Levels of Physical Activity.* London: National Institute for Health and Clinical Excellence. Available at www.nice.org.uk/guidance/ph8.

15 US DOT (2010).

16 Minnesota Life and Securian (2008). The new buzz in health: wellness programs offer savings for employers, a new revenue source for producers. Minnesota Life. Available at www.lifebenefits.com/lb/pdfs/F62382-20%20Get%20More%2012.pdf.

17 US DHHS (2003). *Prevention Makes Common "Cents".* Washington, DC: US Department of Health and Human Services. Available at http://aspe.hhs.gov/health/prevention.

18 Berry, L., Mirabito, A. and Baun, W. (2010). What's the hard return on employee wellness programs? *Harvard Business Review.* Available at http://hbr.org/2010/12/whats-the-hard-return-on-employee-wellness-programs/ar/1.

19 Denning, E. (2011). Exercise in the workplace: part of Surgeon General's vision for healthy and fit nation. *Corporate Fitness Works*, February 18. Available at http://corporatefitnessworks.com/exercise-in-the-workplace-part-of-surgeon-generals-vision-for-healthy-fit-nation/.

20 Walsh, D. (2012). Taking sustainability seriously. *Green at Work*, January 27. Available at http://greenatwork.com/blog/2012/01/27/taking-sustainability-seriously/.

21 Nutramed (2011). Cars, trucks, air pollution and health. Alpha Online. Available at www.nutramed.com/environment/cars.htm.

22 Ibid.

23 US EPA (n.d.) *Clean Air Act.* Washington, DC: US Environmental Protection Agency. Available at www.epa.gov/air/caa/.

24 US EPA (2011). *Benefits and Costs of the Clean Air Act: Second Prospective Study – 1990 to 2020.* Washington, DC: Environmental Protection Agency. Available at www.epa.gov/clean-air-act-overview/benefits-and-costs-clean-air-act-1990-2020-second-prospective-study.

6

GOOD-BYE FRIED CHICKEN, HELLO HEALTHY, SUSTAINABLE FOOD

Patient satisfaction + sustainable foods

FIGURE 6.1 A prescription for health

Executive summary

Promoting community health starts with promoting individual health.[1] One of the most rewarding health promotion tactics a hospital can practice is serving healthful, sustainable food to its patients, visitors and caregivers. Research indicates that nutrition is intrinsically linked to health; nutritious foods support healing and prevent disease.[2] The environmental and financial impacts of using sustainable foods are also worth noting. Locally sourced, fresh produce is frequently the healthiest choice of food on the market, and requires less energy consumption for transportation compared with foods shipped from distant locations.[3] This chapter guides healthcare leadership through the transition from the existing food paradigm to one focused on "healthy" foods in healthcare. Healthy food in healthcare is based on a framework of purchasing local, seasonal foods harvested in a sustainable manner with verified sourcing. Good-bye fried chicken, hello healthy, sustainable food!

A told story

Fred, the Memorial Hospital CEO, receives a letter from a dissatisfied patient, Julie Schmooly, about the food she and her family were given during Julie's hospital stay. The care was great according to Julie, but the food was so bad, Julie is not sure whether she will return to Memorial Hospital should the need arise. Poor food quality detrimentally impacted Julie's patient experience and perceived quality of care. Reflecting upon Julie's letter and his own opinions about hospital food, Fred approaches Liam Ludwig, Memorial's Director of Food Services, about starting a healthy, sustainable foods program for patients, caregivers and employees. Fred asks Liam to partner with the Sustainability Director, Amalie DuBois, on the project.

April 28, 2012

Dear Mr Fred,

I am the proud mother of a baby girl, Louise, after giving birth at Memorial Hospital. I couldn't possibly ask for a better support system. My parents and my husband's parents came to the hospital, along with some of my closest friends and cousins. The nurses were also a wonderful support team. They were attentive to my needs and expressed sincere concern for my well-being; I felt as though I was their only patient, knowing full well that couldn't possibly be true. Dr. Gandhi prepared me well and I felt the delivery went as smoothly as it could have gone.

The food served at Memorial Hospital, however, was just awful, both for me and for my family. If you can't put together a healthy plate of food, what makes me think you can put together a healthy person? I stayed overnight at the hospital, but barely made it on the canned, chemically laden, fatty, frozen foods that came my way. I actually felt bad about having my guests visit the hospital because there wasn't any place to get healthy, fresh meals. My husband ended up going to the nearby McDonald's to get me food, and that's really not what I had in mind for kicking off my post-pregnancy health diet. Have you sampled the food you serve to patients and families?

Mr Fred, I request that you look into alternative food options, food that is healthy and fresh and helps prepare patients to return to daily living and engage in healthy habits.

Sincerely,
Julie Schmooly, Memorial Hospital patient

An insightful conversation

FRED: Liam, I received a negative letter from one of our patients. The patient wrote that her clinical care was excellent, but the food we served her and her family did little to aid her healing process. We know that eating fresh foods and vegetables is one of the keys to staying well and regaining one's health, so maybe we ought to examine our food selection.[4]

LIAM LUDWIG: [*Director, Food Services*] The double cheeseburger and fries combo is on sale today in the cafeteria.

FRED: Are you serious?

LIAM LUDWIG: Yes, I am serious.

FRED: Wow, that's not good, especially since we just launched our "Fight Obesity" campaign. What do you suggest we do about that?

LIAM LUDWIG: Well, I think the hospital has a role in promoting healthy food options. We should be a model of health for our community and our patients. It's just that the cheeseburger sells.

FRED: I understand that. But if we've got patients with congestive heart failure and they're in the hospital for the sixth time, it's probably not a good idea to serve the triple whopper with a week's supply of fat and salt. That'll just make their recovery even harder.

LIAM LUDWIG: Point taken.

Leading the way

The public generally believes that most features of hospitals are the result of intentional decisions by leadership.[5] This is not a happy fact when considering that patients, visitors and employees wonder what people in leadership are thinking when they see the hospital providing onion rings, fried chicken and one slice of pizza drowned in four servings of cheese. One of the most useful functions which sustainability-oriented leaders can provide is to focus on telling the story about sustainability efforts for which there is no debate. Evidence-based practices that positively impact health and the environment should be visibly leveraged with regard to food choices.

The changing wellness paradigm is about individuals taking more responsibility for managing themselves in every dimension of health. Incentives are increasingly offered for employees to improve their diets and enroll in exercise regimens, not to mention that employee health benefits are increasingly discounted for not smoking.[6]

NEW NON-SMOKING HIRING POLICY AT CLEVELAND CLINIC

To take further steps in preserving and improving the health of its employees and patients, Cleveland Clinic implemented a non-smoking hiring policy requiring all job applicants and individuals receiving appointments to take a cotinine test during their pre-placement physical exam. This is a pre-employment test only. The cotinine test detects the presence of nicotine in all forms of tobacco. Beginning on September 1, 2007, appointments offered to prospective residents and fellows who tested positive were rescinded. Individuals who tested positive received a referral to a tobacco cessation program paid for by Cleveland Clinic. Those individuals testing positive who test negative after 90 days may be reconsidered for appointment at the discretion of the program director should the residency position remain vacant.

Source: http://my.clevelandclinic.org/Documents/Urology/Non-Smoking_Hiring_Statement.pdf

---

When employees adopt healthy habits, they achieve reduced health insurance premiums.[7] According to Dr. Dean Ornish, President and Founder of the Preventative Medicine Research Institute,

> Many people tend to think of breakthroughs in medicine as new drugs, lasers, or high-tech surgical procedures. They often have a hard time believing that the simple choices that we make in our lifestyles – what we eat […] can be as powerful as drugs and surgery, but they often are. Sometimes even better.[8]

Wellness and patient health are inherently tied to sustainability.

When behaviors are presented as contributors to a desirable vision, many individuals will embrace the desired new behaviors, even if they have some emotional discomfort associated with behavior change.

Fried chicken tastes great to many people. But a vision of community and health speaks about giving people healthy food options and supporting local agriculture. It is about promoting a healthy childhood, even if the child prefers sugary drinks. Sustainable foods are about preventing cavities and chronic illness in children. There is one inexplicable question for most leaders: why would you promote a smoke-free environment, or ensure everyone gets a flu shot, and then offer unhealthy foods? Effective leaders pay attention to casting the vision of health promotion and tie it to the organization's mission of health. Effective leaders also articulate values, vision and the rationale for change.[9] It is all about making the vision visible and explaining the rationale for the leadership actions taken.

It is easy to tell key constituents that eating fresh and local food is part of the healthcare leader's mission to achieve a healthy community. The good news is that most employees and patients already believe strongly in the organization's vision and values. That shared vision is why many employees will experience some inconvenience and limited restriction in their choices. Employees may be willing to pay a little more for healthy, fresh, local foods. For instance, Whole Foods Market raised its fiscal 2012 outlook after reporting profits of $118.3 million, or 65 cents a share, up 33 percent from earnings during the same period in the previous year (see www.marketwatch.com/story/updates-advisories-and-surprises-2012-02-08). The truth is that leadership and telling the "why" of the vision is important in getting support from stakeholders affected by such changes.

How did we get to where we are?

In the early 1900s, much of America was a rural landscape with food produced and consumed locally.[10] Food was grown at home or in the immediate vicinity, resulting in communities that understood the production process and where food came from. Today, people are generally less connected with their food because food production is often managed by large, global businesses or cooperatives. In fact, two companies are responsible for 50 percent of exported grain; three companies are responsible for 80 percent of the meat slaughtered; and four companies control 90 percent of the cold cereal production.[11] Being removed from the process has greatly reduced the individual consumer/purchaser's awareness of the positive health impacts of food choices, as well as the type and impact of food production methods.[12]

Healthy food tends to be more sustainable fiscally, socially and environmentally.[13] Healthy foods also have a positive impact upon human health. Research shows that, compared with the people who consume the least fiber, people who consume the most fiber have a 40 to 50 percent reduced risk of colorectal cancer. Similarly, those who consume the most vegetables have a 52 percent lower risk of colon cancer.[14] From environmental and fiscal perspectives, food that is purchased locally generally costs less to transport than food purchased from great distances.[15] Local food retains its nutrients better because it is traveling a relatively short distance over a relatively short time period, and can be picked at a time when it is fully matured.[16] In other words, purchasing local foods is the socially responsible, environmentally friendly and healthy thing to do.

Lack of understanding of what makes food healthy and sustainable is a major issue, as evidenced by staggering obesity rates. The Centers for Disease Control and Prevention (CDC) estimate that 35.7 percent of America's population are obese.[17] The cost of treating obesity-related diseases is anticipated to reach $550 billion by 2030.[18] Many healthcare organizations have been slow to recognize the clear connection between their mission of health and providing healthful, sustainable meals. The growing global population and accompanying strains on our environmental resources make a case for highlighting the connection between responsible production and purchasing healthful local food and health.[19] From a fiscal perspective, "Americans spend 9 to 14 percent of their disposal income on food, as compared to about 30 percent in 1950." Based on these statistics, it is natural to wonder whether perhaps the quality of the food purchased helps explain the lower food costs. Researchers and health professionals argue that lower costs do not compensate for the costs of negative downstream health impacts, or the downstream costs of production methods that cause environmental degradation.[20, 21]

Sustainable food is a leadership issue because the average person is not thinking about food as a direct health, environmental or financial issue. Healthy food has a vital place in the realm of individual well-being and is also a central feature of a sustainable model of societal health.

SUSTAINABILITY IN ACTION

Stacia Clinton, RD, LDN, is a Healthy Food in Health Care Program Coordinator for the global non-profit Health Care Without Harm (HCWH). She guides local and sustainable institutional purchasing and program development for the six-state New England region. Nationally, Stacia directs the organization's Healthy Beverage Program and provides technical assistance to healthcare facilities in support of community health and obesity prevention. Stacia has extensive experience in food service and clinical nutrition management. She previously co-owned and directed Nutrition and Wellness Counseling, a successful private practice in Connecticut with a focus on holistic nutrition. Stacia speaks nationally about sustainable food systems and obesity prevention strategies.

How did the work of "Health Care Without Harm's Healthy Food in Health Care Program" start?

> When HCWH started the Healthy Food in Health Care Program ten years ago, there was a complete disconnect in the healthcare sector between healthy food and clinical care. Hospitals thought nothing of having sugar-sweetened beverages, serving large amounts of meat and having fast food in their lobbies.

In 2005 Health Care Without Harm hosted the first Foodmed conference on health and sustainability in hospital food service in Oakland, California. In 2006 The Healthy Food in Health Care Pledge launched nationally, providing a framework for the healthcare sector to support an environmentally sustainable, economically viable and socially just food system. As of 2015, 550 facilities nationally have signed the Pledge (HWH, 2015).

What positive outcomes have you seen emerge from your work?

Our vision – health by way of a healthy food system – is built on the understanding that all aspects of the food system, including how food is grown, processed, packaged, transported and consumed, have implications for the health of individuals, communities and the environment. This system-based approach broadens our sphere of concern beyond individual responsibility and a medical model focused on treating symptoms to promoting health at multiple scales.

We've seen an increase in sustainable and healthy food purchases by the healthcare sector, increased efforts to build a healthier food system, and increased clinical advocacy around important public and environmental health issues, such as antibiotic resistance. We've also seen increased collaboration between institutional sectors to build a healthier food system together.

How has the sustainable food marketplace within healthcare changed since you began working in the field?

Healthcare has become much more aware that its food purchases have impact beyond feeding staff and patients. Awareness of the therapeutic value of food has increased as healthcare embraced the concept "food as medicine." Healthcare is now considering the health properties of food beyond nutrients to reflect on other chemicals we may be consuming, as well as agricultural practices and resulting public health impacts hospitals are promoting through their purchases. In some ways, we are coming full circle. Years ago, the healthcare sector used more whole foods and cooked from scratch. In favor of efficiency and reduced labor costs, new processed foods became popular.

The new-found attention to the therapeutic quality of food in treating patients or preventing health issues in hospital staff has hospital food services returning to more whole, fresh foods, particularly those that may be sourced from local communities. However, I suspect that if you were to ask a food service director, she or he would tell you that the institutional food supply chain is lagging behind demand as large distributors, food service management companies and group purchasing organizations are slow to prioritize local and sustainable food sources.

Explain why meat from animals raised without antibiotics is better for our health than the alternative. How and why are hospitals currently making antibiotic-free meat a priority for their food services? How difficult a task is it for hospitals to serve antibiotic-free meat?

The concern about antibiotic overuse in animal agriculture and the impact upon human health is not about meat consumption. Animals given antibiotics are tested for levels of antibiotic residues and measured against a threshold determined by the FDA. The concern is related to the development of resistant bacteria as a result of the overuse of antibiotics in conventional animal agriculture. According to government estimates, more than four times the amounts of antibiotics are sold for use in animal agriculture (close to 30 million lb) compared to human medicine. Antibiotics are routinely given to poultry, beef cattle and swine to prevent disease outbreaks that are inevitable under confined, unhygienic conditions. These are the same antibiotic classes used to treat human infections, including penicillins, tetracyclines, sulfa drugs, macrolides and more.

There is a strong consensus among health experts that current rates of antibiotics in animals pose a threat to human health by increasing the development of antibiotic-resistant bacteria. It is a position shared by the U.S. Institutes of Medicine (IOM), the U.S. Centers for Disease Control (CDC), the World Health Organization (WHO) and leading medical associations. This reality has become a major threat to the US population's health because we now have greater exposure to resistant bacteria through the handling of meats as well as through contamination of our waterways, air and soil. In particular, individuals who work on or live in close proximity to farms with overuse of antibiotics have even greater rates of infection.

Hospitals are prioritizing antibiotic resistance in meats because they are on the frontlines dealing with the effects. Antibiotic resistance increases the number of bacterial infections, increases the severity of those infections, and helps increase hospital costs. In the United States, close to 19,000 deaths from methicillin-resistant Staphylococcus aureus (MRSA) occur on an annual basis. Longer, more expensive hospital stays for treating antibiotic-resistant infections cost the healthcare sector $21 to $34 billion each year.

As stewards of antibiotics, doctors and hospitals have created rigorous guidelines to end the overuse of antibiotics in human medicine. Now, forward-thinking hospitals are expanding their roles as stewards of antibiotics by using their purchasing power to support producers that use antibiotics responsibly, and by pushing for strong public policies that will protect antibiotics.

Since we began assisting hospitals in sourcing meat raised without the routine use of non-therapeutic antibiotics, we have seen increased availability of non-therapeutic antibiotics and associated decreased health costs. Today, an additional cost exists to transition purchases to these meat products because meat production without routine antibiotics is still not widespread. When compared to the costs associated with treating antibiotic-resistant bacterial infections, the cost of treatment is exponentially higher than the additional costs to transition meat products. Hospitals continue to struggle to identify sources of these meat products on the approved order guides of their food service management companies. Some hospitals find that sourcing closer to home from small or mid-scale producers whose production is often scale-appropriate and does not require prophylactic antibiotic use has been easier than trying to change the purchasing habits of national supply chain entities.

What tools or resources have you found helpful in your journey toward sustainable healthcare?

The resources we create at Health Care Without Harm and Practice Greenhealth continue to be cutting edge, and primary resources for guiding the healthcare sector toward sustainability. In compiling those resources we rely on vast amounts of research, including peer-reviewed journals and other practice-based research, as well as hands-on interactions with producers, including fishermen, and supply chain entities. In addition, we support the development of hospital focus groups to test strategies and pilot innovative programs that improve education and accessibility of local and sustainable food products. Research from the Johns Hopkins Center for Livable Future and the Michigan State University Center for Regional Food Systems, among others, has been especially useful regarding the impact of agricultural production practices upon our environment and the subsequent connection to human health.

What is your advice for latecomers to sustainability?

We all start somewhere. One place to start is by learning best practices through literature and by speaking with peers who have been engaged in the sustainability field for some time. Another solid first step is to understand the link between agricultural production practices, the health of our environment, and the health of people and communities. You may consider touring a conventional farm as well as an organic farm to understand the differences. Taking any step, no matter how small, is a step forward.

UNDERSTANDING LOCAL VERSUS SUSTAINABLE

There is sustainable, and then there is local. Local food is often defined as being sourced within a 100- to 400-mile radius.[22] On the average American food plate, food comes from about 1,500 miles away[23] (CUESA, n.d.) That is why the concept "buy fresh, buy local" is a powerful two-for-one deal in terms of impact[24] – the "buy local" is about supporting our local economy in ways that promote minimal environmental degradation. And the "buy fresh" is about fresh foods – it is the alternative to food with lots of sodium often found in canned vegetables, or packaged with preservatives or sugar. Dishes with plants and fruits are most sustainable because they reduce environmental damage and have a positive public health impact.

Why healthy, sustainable foods matter to healthcare

US hospitals were not traditionally paid for keeping people healthy.[25] That is why most hospitals and most health professionals are paid to treat sick people. Frankly, it was often outside most health providers' economic interests to attempt to promote healthful behaviors. The old cliché that "care follows the dollar" is true in many circumstances. But with the incentives provided by many insurers to penalize hospitals and doctors for unnecessary readmissions, times are changing, and so, too, are attitudes. Hospitals need to feed patients healthy, sustainable foods so that they do not return to the hospital. Part of that new responsibility includes teaching patients not to go across the street to a fast-food restaurant for high-fat, high-calorie food when they leave the hospital.

All that fried chicken is like fins on a 1957 Chrysler. That was the model feature that made perfect sense at the time. Today, dealers find that fins have extra weight and, with mileage going down and the high cost of gasoline, fins are a problem. Similarly, fried food is making people sick and healthcare will increasingly be paid to keep people healthy. Most hospital leaders have come to understand that they are trapped in a model that will eventually collapse precisely because it focuses too much on sickness instead of on health.

A changing paradigm

By 2030 40 percent of the US population is anticipated to have some form of cardiovascular disease (Statin Usage, 2013). The potential cost savings of reducing the need for statin drugs by changing diet and lifestyle is of great interest.[26] Indeed, a shift in public attitude is underway. Healthcare is transitioning to a new paradigm in which the goal is to focus on health and well-being, preventing people from needing to enter hospital until absolutely necessary.[27]

The new mantra recites healthcare's obligation to have a positive impact upon individual health. There is some disagreement in the policy debate about the connection between sustainability and health promotion, but the problem is that

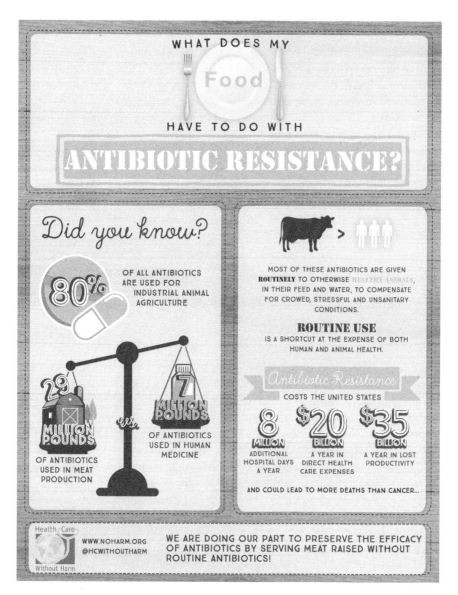

FIGURE 6.2 What does my food have to do with antibiotic resistance?

everything from housing to clothing to education to public safety impacts health.[28] This chapter helps focus the scope of health on food, a key ingredient of individual, family and community well-being. What is the role of leadership in helping patients, families and health professionals understand the critical role that sustainable, healthy food has in optimizing the individual's health status and quality of life?

To begin with, leaders must encourage individuals and the healthcare system at large to think differently about food. Healthy food is part of our mission and our responsibility to model the sustainability and healthful character of our industry.

A leadership dilemma, challenge or opportunity?

Healthcare administrators at best practice organizations are influencing food vendors to make different selections when it comes to food services. Best practice organizations such as Kaiser Permanente, Cleveland Clinic and Vanguard Health Systems host farmers' markets and engage physicians in promoting healthy, local food habits for their patients. They eliminate sugary drinks and encourage serving healthy foods in the cafeteria, among other food strategies.

Kaiser's and Vanguard's sustainable food practices deserve praise, but what about the rest of the healthcare industry? Many organizations are advocating for a balanced menu to be implemented immediately, which means increasing the average amount of fruit and vegetables on the menu of healthcare organizations.[29] This is just one example of a small but meaningful step that healthcare leaders can take to embrace sustainable foods and healthful living.

Healthy food, healthy economics

> One quarter of what you eat keeps you alive. The other three quarters keeps your doctor alive.
>
> *(Egyptian saying)*

One dilemma in leadership is the challenge of balancing customers' desires with the vision and mission of the organization. The vision of a healthy, sustainable community and happy, satisfied individuals sometimes conflicts with the habits and perceptions that both the community and patients embrace. For instance, some hospitals pride themselves on offering whichever foods patients request. The problem with this approach is that when hospitals ask patients, "Would you rather have the tofu patty with steamed broccoli or a double order of fried chicken with a side of buttered potatoes?," well, you know what people will pick. That is why so many patients end up in the hospital in the first place, with a variety of illnesses linked to obesity.[30]

At face value, the opt-in is the higher purpose – health and well-being, or the short-term pragmatic gain of a happy fried chicken customer. If sustainability and health are two of healthcare's higher purposes, then leaders have to make these priorities visible and explicit. Leaders make priorities clear and then align leadership around these priorities by communicating them throughout the organization. Leaders who succeed will be clear about the role that healthy, sustainable foods play in keeping employees healthy and helping patients optimize their health and the health of their families.

Dear Julie,
Thank you for bringing to our attention the difficulties you encountered during your visit at Memorial Hospital. As always, our team strives to provide the highest quality care to our patients and we truly appreciate your feedback. We view your concern as an opportunity to improve the services we provide.

Since receiving your feedback, I met with Memorial Hospital's leader of Food Services to discuss why and how to integrate sustainable foods for patients, employees and visitors alike. Memorial Hospital now has a plan to ensure that food is part of how we provide excellent care for future patients.

Should you be in the area again soon, we'd be delighted if you'd come visit our new onsite farmers' market that supports local agriculture while promoting healthy, fresh eating habits or take some of our new cooking classes focused on healthy sustainable foods. We're also in the process of revamping our cafeteria dining options to incorporate and incentivize employees and visitors to consume more healthful foods. Patients, similarly, will have access to foods that optimize health and well-being.

Again, we are truly sorry that you were not completely satisfied with your visit to Memorial Hospital and we hope that you will continue to use our facility for your family's future healthcare needs. It is through feedback like yours that improvements are made throughout Memorial Hospital.

Sincerely,
Fred
Chief Executive Officer, Memorial Hospital

FIGURE 6.3 Hospitals as anchor institutes

Practical application: eliminating sugar-sweetened beverages in healthcare

The imperative

- Foster a culture of health and wellness within healthcare organizations.
- Fulfill organizational mission related to improving the health of patients, employees and communities.
- Reduce rates of obesity and chronic illness as well as their associated healthcare costs.
- Increase public and community goodwill towards the organization.
- Capture financial incentives of healthier foods programs, including national grant opportunities.
- Create and capture opportunities for publicity.

The initiative

Healthcare organizations across the United States are working with government and non-government agency (NGO) partners to develop hospital food service programs that promote health and wellness. A major focus of these efforts has been to reduce and/or eliminate sugar-sweetened beverages (SSBs) within healthcare facilities, since this type of beverage has been shown to be detrimental to public health.

Several industry-leading healthcare organizations are acting on this mission, each at a different point along the journey to eliminate SSBs within their walls. They are targeting all beverages to which a caloric sweetener is added, including sugar, high-fructose corn syrup, cane sugar, glucose, sucrose, honey and brown sugar, among others. Beverages on the target list include sodas, sweetened tea and coffee drinks, sports drinks, energy drinks and fruit juices.

Programs to eliminate SSBs often consist of a phase-out schedule, which generally includes the following steps:

1   Create a project task force with representatives from multiple departments, including food services, patient care, leadership, etc.
2   Conduct a beverage audit to identify availability and consumption levels of both SSBs and public drinking water within each facility.
3   Develop a healthy beverage program outline with input from the project task force (the outline should identify specific action items such as removing SSBs from vending machines, adjusting pricing structures to incentivize healthier beverages, increasing visibility of non-SSBs, etc.), create a timeline for achieving the goals, and establish accountability for outcomes.
4   Adopt a healthy beverage policy that includes purchasing specifications for future contracts.

Initiatives aiming to eliminate SSBs from the healthcare setting are often part of an organization's healthy foods program and healthy beverage consumption habits. Additional components of such programs may include efforts such as:

- Eliminate fried foods.
- Expand salad bar offerings.
- Replace unhealthy options with healthier choices in vending machines and cafeterias.
- Develop a pricing structure that makes healthy choices more financially attractive.
- Increase labeling and education efforts in locations where food is sold.

The metrics

Measuring progress toward achieving the elimination of sugar-sweetened beverages may be achieved by tracking the following:

- Total sales of SSBs in cafeterias, vending machines, catered events, etc.
- Total availability of SSBs in cafeterias, vending machines, catered events, etc.
- Placement of SSBs versus non-SSBs in most visible retail locations.
- Number of public water sources (with or without filters).
- Results of food service surveys regarding attitudes, knowledge and opinions about SSBs.

Lessons learned

- Do not be alarmed by initial drops in sales: several organizations that have implemented a complete ban on SSBs saw drops in sales after initiating the ban, but later recouped in full any initial losses experienced.
- Increasing public access to free water – and ensuring that water is appealing to consumers – is a crucial component of shifting beverage consumption habits. Try adding filters to existing water machines, installing additional filtered water machines, creating signage and educational materials, providing reusable water containers, or financially incentivizing the consumption of water over SSBs.
- Several steps may be taken to make the consumption of healthier beverages a no-brainer for consumers, including making healthy beverages cheaper than SSBs, adjusting the amount and location of shelf space given to SSBs, highlighting healthier beverage options, and creating educational materials highlighting the health and environmental impacts of high consumption of SSBs.
- Facilitate implementation at the procurement end by establishing a general contract with beverage vendors. Such contracts will not specify a percentage of SSBs sold, so the facility is able to select which beverages it wishes to offer.
- Before taking any action, it is crucial to gain buy-in from hospital administration and clinical advocates. This level of backing creates an environment of social support and will help head off any potential pushback from consumers.
- Education and communication are critical when implementing initiatives that require consumers to change their behavior patterns. Ensure staff and other consumers have adequate time and mechanisms to provide feedback before rolling out the program. Creating messaging that highlights the health impacts of the initiative will also help facilitate buy-in.

About the organization

Healthcare organizations such as Vanguard Health Systems, Cleveland Clinic and Fairview Hospital are leading the way nationwide in the shift toward healthier beverages in healthcare. They are supported by initiatives such as Health Care Without Harm's Healthy Beverage Project and the CDC's Rethink Your Drink campaign, among others.

Sources

CUESA (n.d.). How far does your food travel to get to your plate? Available at https://cuesa.org/learn/how-far-does-your-food-travel-get-your-plate (accessed 10 August, 2017).

Healthcare Without Harm (2015). *Menu of Change – Healthy Food in Healthcare* (pp. 1–32, Rep.). Reston, VA: Healthcare Without Harm. Available at https://noharm-uscanada.org/hcwh-content-tags/menu-change-2015 (accessed August 6, 2017).

Statin Usage (2013). The cost of high cholesterol. Available at www.statinusage.com/Pages/the-cost-of-high-cholesterol.aspx (accessed August 6, 2017).

http://articles.chicagotribune.com/2012-04-24/business/ct-biz-0425-hospital-sugar-ban-20120424_1_sugar-sweetened-beverages-drinks-hospital-group/2.

www.chicagohispanichealthcoalition.org/.

https://noharm-uscanada.org/lib/downloads/food/Hydrate_For_Health.pdf.

https://noharm-uscanada.org/documents/fairview-hospital-healthy-beverages-case-study.

Leadership imperative

One goal of the inpatient care team is to prepare a patient and their accompanying family to resume their lives and maintain health once they return home. In the hospital, patients are surrounded by nutritionists, nurses, pharmacists and therapists; it is like football training camp where there are many expert coaches surrounding the players before they actually play the game. In the game, there is no coach on the field, there is no weightlifting guru, there is no nutritionist. Likewise, when one is an inpatient, it is training camp, it is practice: the hospital tries to build a model that bridges between the patient and an independent community citizen. At home is real life. The way to keep score is the quality of health.

When a patient and family are in the hospital, they have the opportunity to be explicitly educated or "coached" in terms of what to eat, how to prepare healthy food and how to select foods as part of building their health. The foods hospitals serve patients, the ways hospitals educate families about healthy, sustainable food habits, the tools hospitals give patients and families about healthy eating – when and how much – are all about optimizing individual and community health. The incentives to keep people healthy have been coming for a long time. Now the incentive is changing: good-bye fried chicken, hello sustainable foods.

Leadership dialogue

1   What strategies can healthcare leaders use to educate employees and caregivers on the role a healthful diet plays in preventing disease and speeding recovery?
2   Lack of knowledge about the importance of sustainable food in both patient and employee settings can lead to resistance when introducing healthier, more sustainable alternatives. What can leaders do to overcome such resistance?
3   One of the most difficult cultural elements to change is persuading patients to be accountable for their health and food choices. How should leaders support physicians in their efforts to encourage patients' healthful, sustainable food choices?
4   Community leaders increasingly encourage hospital leadership to "walk the sustainability talk." Why should or shouldn't hospitals continue to offer sodas and sodium-rich snacks in their visitor vending areas?
5   A patient's time spent in the hospital is time spent preparing for "real life" following discharge. How should leaders encourage employees and physicians to empower patients to select, prepare and expect healthy food choices after they return home?

Syllabus integration

Overview

Students will assess and analyze environmental issues related to local, nutritious food procurement in the healthcare setting.

Review case studies that demonstrate how the current healthcare food supply and distribution system can be modified to comply with sustainability aims.

Objectives

1   Identify factors that pose a risk to the operations of daily sustainable food procurement:

a   Food safety objectives.

2   Explore incentive programs structured to advance sustainable food objectives, including local procurement and healthy eating habits:

a   Menu of change.

3   Identify the link between human health and food systems.

Topics

*   Food sourcing:

a   Farmers' markets and nutrition
b   Carbon footprint
c   Green chemistry.

- Case studies:

  a Pitt County Memorial Hospital
  b Children's Hospital and Regional Medical Center, Seattle.

Homework

Healthier Hospitals (2017). Healthier foods overview. Available at http://healthierhospitals.org/hhi-challenges/healthier-food.

Health Care Without Harm (2017). Healthy food in health care. Available at https://noharm-uscanada.org/issues/us-canada/healthy-food-health-care.

Texas Department of State Health Services (n.d.). Farm to work toolkit resources. Available at www.dshs.state.tx.us/obesity/pdf/F2W_toolkit_resources.pdf.

Submit a one-page proposal to the Board of Directors outlining a proposal for sustainable food procurement. Record in your career journal.

Additional resources

Healthier Hospitals (2017). Healthier foods case studies. Available at http://healthierhospitals.org/challenge/healthier-food.

Practice Greenhealth (2009). Greening operations series: food services waste prevention and composting. Webinar, January 16. Available at www.practicegreenhealth.org.

www.nems-v.com/NEMS-VTools.html.

www.cdc.gov/nccdphp/dnpao/hwi/programdesign/wellness_committees.htm.

http://healthierhospitals.org/hhi-challenges/healthier-food.

www.ncpreventionpartners.org/dnn/.

Notes

1 bryandamis (2011). A holistic approach to environmental public health. *National Environmental Health Promotion Network*, September 29. Available at nehpnblog.com/?p=73.

2 National Prevention Council (2011). *National Prevention Strategy: America's Plan for Better Health and Wellness*. Rockville: National Prevention, Health Promotion and Public Health Council. Available at www.surgeongeneral.gov/priorities/prevention/strategy/report.pdf.

3 Oglethorpe, D. (2008). Local food – miles better? *World Watch* 22(3): 12–15. Available at www.europeansupplychainmanagement.co.uk/article-page.php?contentid=4203&issueid=174.

4 WebMD (2009). Food & recipes: healthy eating. *WebMD*, October 12. Available at www.webmd.com/food-recipes/tc/healthy-eating-overview.

5 Hyde, L. (2011). "Confessions of a formal education enabler." *Museum Commons*, August 8. Available at http://museumcommons.blogspot.com/2011_08_01_archive.html.

6 Hand, L. (2009). Employer health incentives. *Harvard School of Public Health*. Available at www.hsph.harvard.edu/news/magazine/winter09healthincentives/.

7 Ibid.

8 Ornish, D. (2009). Mostly plants. *American Journal of Cardiology* 104(7): 957–8. Available at www.ajconline.org/article/S0002-9149(09)01098-4/abstract.

9 Whitlock, J.L. (2003). *Strategic Thinking, Planning, and Doing: How to Reunite Leadership and Management to Connect Vision with Action*. Washington, DC: George Washington University Center for Excellence in Municipal Management, mimeo.

10 Why local? *Get Local Foods*. Available at http://getlocalfoods.com/Why-go-Local/The-Issue.aspx.

11 Krebs, A. (n.d.). Corporate agribusiness: monopolising subsistence. *Converge*. Available at www.converge.org.nz/pirm/corpag.htm.

12 Wickel, R. (2012). From class to farm: students grow understanding of food production. *The Pendulum: Elon University's Student News Organization*, May 1. Available at http://elonpendulum.com/2012/05/from-class-to-farm-students-grow-understanding-of-food-production.

13 Brown, D., Berko, P., Dedrick, P., Hilliard, B. and Pfleeger, J. (2010). *Burgerville: Sustainability and Sourcing in a QSR Supply Chain*. Portland, OR: Center for Global Leadership in Sustainability, School of Business Administration, Portland State University. Available at http://pdxscholar.library.pdx.edu/cgi/viewcontent.cgi?article=1032&context=busadmin_fac.

14 Campbell II, T. and Campbell, C.T. (2008). The benefits of integrating nutrition into clinical medicine. *Israel Medical Association Journal* 10: 730–2.

15 Sustainable Table (2009). Why buy local? *Sustainable Table*. Available at www.sustainabletable.org/issues/whybuylocal/#fn8.

16 Frith, K. (2007). "Is local more nutritious?" It depends. Center for Health and the Global Environment, Harvard Medical School. Available at www.chgeharvard.org/sites/default/files/resources/local_nutrition.pdf.

17 CDC (2012). Overweight and obesity. Centers for Disease Control and Prevention. Available at www.cdc.gov/obesity/data/facts.html.

18 Sabrina, A. (2010). America's obesity epidemic to cost $550 billion by 2030: study. *International Business Times*, May 8. Available at www.ibtimes.com/articles/338553/20120508/america-s-obesity-epidemic-hangs-550-billion.htm.

19 Parker-Pope, T. (2008). Boosting health with local food. *New York Times Well Blogs*, June 6. Available at http://well.blogs.nytimes.com/2008/06/06/boosting-health-with-local-food/.

20 Drewnowski, A. and Darmon, N. (2005). Food choices and diet costs: an economic analysis. *Journal of Nutrition* 135(4): 900–4. Available at www.ncbi.nlm.nih.gov/pubmed/15795456.

21 Ornish, D. (2012). Holy Cow! What's good for you is good for our planet. *Archives of Internal Medicine* (online first) E9–E10. Available at www.natap.org/2012/newsUpdates/archinternmed.2012.174v1.pdf.

22 Martinez, S., Hand, M. and Pollack, S. (2010) *Local Food Systems Concepts, Impacts, and Issues*. Washington, DC: US Department of Agriculture. Available at www.ers.usda.gov/publications/pub-details/?pubid=46395.

23 Sustainable Table (2009).

24 Buy fresh buy local. *Food Routes*. Available at http://foodroutes.org/buy-fresh-buy-local-program/. jsp.

25 Flower, J. (1993). Getting paid to keep people healthy: two ways of integrating a health-care system. *Healthcare Forum Journal* 36(2): 51–6.

26 Ornish, D. (2009).

27 Lipson, D.J. and Simon, S. (2010). Quality's new frontier: reducing hospitalizations and improving transitions in long-term care. *Mathematica Issue Brief* 7. Available at https://ideas.repec.org/p/mpr/mprres/307fb0f0db884e1699dd1045a1319c2d.html.

28 Mikkonen, J. and Raphael, D. (2010). *Social Determinants of Health: The Canadian Facts*. Toronto: York University School of Health Policy and Management. Available at www.unnaturalcauses.org/assets/uploads/file/The_Canadian_Facts.pdf.

29 Balanced menus initiative. *Health Care Without Harm*. Available at https://noharm-uscanada.org/issues/us-canada/balanced-menus-initiative.

30 Fauntleroy, G. (2007). Obesity leads to more hospital admissions, longer stays. *Health Behavior News Service*, December 11. Available at www.cfah.org/hbns/archives/getDocument.cfm?documentID=1631.

7

THROUGH THE LOOKING GLASS

Cost + energy management

FIGURE 7.1 Hospital energy options

Executive summary

Healthcare ranks second among industry sectors in energy consumption.[1] Hospitals examining ways to manage costs will find that energy projects provide new opportunities. This chapter highlights why and how hospitals seek to reduce operating costs through energy efficiency and energy reduction while maintaining a high quality of care in existing clinical programs.

Storyline

The Chief Medical Officer, Dr. Hedge, and the Director of Facilities, Sandy Smith, are waiting in the cafeteria line. Dr. Hedge turns to Sandy Smith and initiates a conversation.

DR. HEDGE: I can't believe it. We just received our monthly CMS [Centers for Medicare and Medicaid Services] quality measures report, and we're below the midpoint again. Every time this happens, we get calls from the health reporters asking why clinical quality is suffering.

SANDY SMITH: You think that's a problem! We just got a letter from the electric company letting us know that we're one of the largest energy users in the entire state. Just wait until our elected officials get a load of that information and decide to do something with it!

DR. HEDGE: Well, Sandy. [...] That actually makes a lot of sense. After all, we do so many CT scans, and we do a lot of MRIs too. Let's face it, the tests cost a ton of money and use an immense amount of energy.

SANDY SMITH: Well, Dr. Hedge, the last time I looked, your CT scans were my number one energy consumer during the 9 to 5 shift. [*Phone makes a "ping" noise, indicating that a new email message has been received. Sandy Smith checks his phone and sighs.*] Memorial Hospital just got listed in the lowest category in the governor's energy efficiency list. We're in trouble now.

WHY LEADERS DO WHAT WE DO

When you started out as a healthcare leader, no one asked you to think through sustainability practices as part of your leadership responsibilities. So let us pause and ask you now: why do you do what you do? Why do you have all of the lights on in your office, especially in the parts of the office that you are not in most of the time, like the kitchen or the print room? Why do you leave the lights on when you are gone for half the day? Why is the computer on when you are absent for extended time periods? While we are at it, why don't we ever empty the recycling container? Most of our behavior is not intentional; we typically do what we have always done, mostly because we are creatures of habit.[2] Effective leaders are intentional about their personal behavior as a strategy to align leadership teams and to communicate desirable change. Modeling is one way of making desired behaviors visible and instigating change.

WHO TURNED OFF THE LIGHTS?

In a typical healthcare facility in the US, lighting represents 42 percent of electricity consumption, not including its impact upon cooling loads. Posting 'Please turn the lights off when not needed' stickers above light switches will remind both staff and visitors to do so.

Source: http://bea.touchstoneenergy.com/sites/beabea/files/PDF/
Sector/Hospitals.pdf

Modeling desirable behavior is the role of the leader

DR. HEDGE: I think the vast majority of behavior is imitative. It's not rational, thought-driven behavior. Why do you wash your hands instead of using hand sanitizer? Or why do you use hand sanitizer instead of washing your hands? We don't typically have a good answer to those questions. I tell people that 90 percent of personal development will come from simply emulating successful people. It won't be an intellectual exercise. If you're hanging out with people who eat fried foods, what do you think you're likely to eat? The easiest thing to do is eat what everyone else is having, so I guess I'll have fries and pizza. That's not the way we like to think about ourselves, though. We like to think we leaders are so independent and intentional. But I don't care how smart or how driven you are, imitating the behavior of others is the way most people act. So, in truth, picking who you are with and who you work for and respond to is your practical destiny.

SANDY SMITH: How can leaders use that thought process to effect change? Is there a way that leaders can leverage the sustainability mindset to drive change? Perhaps by modeling sustainable behaviors?

DR. HEDGE: That makes sense to me.

SANDY SMITH: If turning off your lights became habitual and everybody slowly started to witness the change, conserving energy would become the norm. You know, Dr. Hedge, you send messages all the time that you may not give much thought to, messages that are purely based on your behaviors, your habits. Those messages could be about sustainability or anything else.

DR. HEDGE: You're right, Sandy. I unwittingly send a lot of messages based on my habits. Leaving the lights on is what I tend to feel most comfortable doing. I do it day-in and day-out, and I don't really think about the message I'm sending because I never bother to ask why I leave the lights on in the first place. I suppose I leave the lights on because it's the easiest thing to do and, quite frankly, I never before gave much thought to sustainability as a leadership imperative. There's certainly a positive power associated with being a role model for good.

How do you lead if you don't know anything?

Leaders often assume the mindset of "I'm transferring some of my superior knowledge to my subordinates or team members I lead. I'm going to determine how we do things around here." But the successful leader eventually realizes that he or she does not know as much as the team members closer to the action. As the leader communicates with her team, she is educated by her team.

DR. HEDGE: You're educating me, and I'm learning as you give examples of energy management – I knew nothing about energy management when we started this conversation. And the thing is, I didn't know I didn't know. Well, I kind of knew I didn't know anything about sustainability.

SANDY SMITH: You probably know more than you think you do. Do you recycle at home? Do you use glasses or Styrofoam cups at your house? Do you leave your home kitchen lights on overnight?

[*Fred joins their table.*]

THE POWER OF REDUCING POWER

According to the US Environmental Protection Agency (US EPA) program:

- Every dollar a nonprofit healthcare organization saves on energy is equivalent to $20 in new revenues for hospitals or $10 for medical offices.
- Some for-profit hospitals, medical offices and nursing homes can boost earnings per share by a material amount by simply reducing energy costs by 5 percent.[3]
- According to the World Health Organization, in the USA the health sector's use of electricity adds over $600 million per year in direct health costs and over $5 billion in indirect costs.[4]

WHAT IS ENERGY STAR?

Energy Star is a joint program of the US EPA and the US Department of Energy. The joint program helps Americans save money and protect the environment through energy-efficient products and practices. In 2010, Energy Star helped Americans save about $18 billion on utility bills in an effort to avoid greenhouse gas emissions.[5]

According to the Department of Energy, reducing the energy use of healthcare facilities offers several benefits:[6]

- Improved profitability
- Reduced impact of volatile energy costs
- Lower operations and maintenance costs
- Improved environmental performance
- Reduced carbon footprint
- Healthier healing and work environment
- Healthier communities.

Putting it in perspective

FRED: It's the notion that I have the power to model sustainable practices because a bunch of people work for me, and they want to mimic the behaviors that I model. That's just the way it works, right? If I don't pause for that period of purposeful introspection before I act, I may well be an unintentional negative influence on the people I lead.

SANDY SMITH: I guess most of the time it's negative messages that are sent by leaving the lights on. If I walk around here, I assume we're not concerned about energy costs at Memorial Hospital because we keep the lights on overnight, and we leave the lights on in rooms where people aren't even present. We don't have any signs posted asking people to shut the lights off when they exit the kitchen or the restroom. So I'm making a negative assumption, based on the facts I'm presented with.

DR. HEDGE: I see.

SANDY SMITH: You know, Fred, when I got my computer set up, I had my reusable water bottle on my desk – the one you call "that canteen thing." And the IT specialist came to help me set up my work station. I had two computers at my desk: one that was a desktop monitor and the other was a laptop. I told the IT specialist that I only needed the laptop and he could take away the monitor. And the IT specialist replied, "Oh, I get it, you're conserving energy." And I said, "Yes, well, I guess that's true." That wasn't my intention, to be completely honest. I just didn't want all that electronic clutter on my desk. So, you know, it was a completely unintentional, positive message that I gave the IT specialist because when she looked and saw the reusable water bottle on my desk, she assumed all of my behaviors embodied a sustainable underpinning.

DR. HEDGE: In your case, you sent a positive message. That's not often the case [*chuckling*].

SANDY SMITH: Right. I think unintentional positive messages are sent a lot less frequently than the negative messages that get articulated. Your position in the organization influences how much your habits are studied. That's why it's so important for leaders to practice sustainability habits: because when the leader demonstrates sustainable behavior, others follow.

"DID YOU KNOW?"

A typical 200,000-square-foot (ft^2), 50-bed hospital in the US annually spends $680,000 – or approximately $13,611 per bed – on electricity and natural gas (2013).

The average cost of power per ft^2 for hospitals in North America is approximately $2.84 for electricity and $0.94 for natural gas (2013).

Source: http://bea.touchstoneenergy.com/sites/beabea/files/PDF/Sector/Hospitals.pdf

---

Make the link explicit between wasted energy and improved work processes

Operational processes should be designed to produce as little unintended by-product as possible. The more people think about waste as a general concept, the better for process improvement and the better for sustainability. We can waste energy, we can waste physical resources, we can waste intellectual capabilities and we can waste creativity. Embedding the concept of waste and the processes that result in sustainable operations maximizes leadership in all dimensions – leaders cannot engage in any meaningful area of their work without encountering the principles of sustainability when thought about in this way.

Lean thinking, process improvement, sustainability – call it what you will – we are using a variety of words for the same idea. It is the notion that we need to look for efficiencies in a holistic manner while tackling our business goals.

YOU CANNOT MANAGE WHAT YOU CANNOT MEASURE

Success can be hard to define, especially if the parameters to measure progress are unclear. Given the fluctuations and general trend of increased cost for energy use, energy efficiency offers an untapped potential for savings. The Energy Use Index (EUI) is a commonly used indicator which is the total amount of energy used by a building per square foot on an annual basis.[7] The EUI is a tool for understanding where energy management opportunities exist and for establishing measurable goals based on industry benchmarks.

FRED: We're very early in that process. Sure, we have a Director of Sustainability, a content expert, but our challenge is that we're relatively immature about integrating sustainability into the organization. My goal is for nurses to think about sustainability as part of their daily routine, the same way that the average nurse thinks about clinical quality or safety. Today, the average

FIGURE 7.2 Benchmarking energy use

healthcare employee doesn't know much about sustainability or what it means to work for a sustainable hospital. We need to train our people in Lean thinking and the sustainability mindset to drive waste out of everything we do. We need to integrate sustainability with our key organizational goals like cost reduction.

DR. HEDGE: The key for us is that we need to make that link explicit. A lot of people don't make the link at all.

FRED: Once upon a time, no one gave much energy – no pun intended! – to the ongoing cost of ignoring sustainability leadership. I wonder: what does wasted energy cost anyway?

DR. HEDGE: I thought you might ask that question. After all, you're going to have to answer it for the Board when we ask them for capital investment funding. Can't say I envy you. […]

HOW MUCH DOES A HIPPO WEIGH?

As hospitals reduce their energy impacts and save dollars in the process, the way the story is communicated plays a critical role in the success of continued employee engagement and public relations success. The average stakeholder will not have an understanding of energy use terminology such as kilowatt hours and carbon tons. Highlighting the success in relation to known parameters – such as number of cars removed from the roadways, or energy use saved in relation to a number of houses or how many pounds of waste in "hippos" they have saved – tells a more effective story.

Energize your leadership priorities

The reality is that energy management has not always been on the leadership's radar – if it gets picked up at all.

FRED: Well, energy is definitely on my radar. When I look at things like natural gas or oil, those energy sources have tremendous variability – I bet in the past five years there's been 100 percent or maybe 200 percent variation. You know what natural gas costs today? Oil consumption has gone way down in total, but the price has gone way up, and I don't understand why that is. Four years ago, it was about 50 bucks a barrel. Now it's about 100-something.

SANDY SMITH: Fair enough, but the cost of natural gas is only a few percent of the total operating budget of a hospital. Large sums of money, but small percentage. The percentile of the total costs probably hasn't changed very much. So why is energy management important now? Is it important because we're in a different economic state?

FRED: It probably has more to do with the state of our national security. We live in a country that's dependent on foreign energy sources, so politics, wars and natural disasters bring on spikes in the cost of energy.

SANDY SMITH: Yes. And so my point is that we can insulate Memorial Hospital from all that as best we can. We can reduce our dependence on volatile energy input prices and insulate ourselves, to some degree, from the variation in the energy cost structure. If we're curtailing our energy use and cutting our cost structure, we know that, down the road, we'll be in a better spot financially than we would be otherwise. That strategy would make our cash flow more predictable.

FRED: Wouldn't that be nice! We have a general sense of where our reimbursements will be headed over the next few years. With energy, though, we don't have that security because it's so volatile. [*Phone rings. Fred checks to see who's calling on his cell phone and decides not to answer.*]

Hospitals are hogs

Today's young professionals were raised with an appreciation for recycling and other sustainable lifestyle choices.[8] In fact, a recent study highlighted that two-thirds of millennials would be willing to give up social media for one week if the entire organization committed to recycling (Recycling Today Staff, 2017). As incoming generations enter the workplace with this mindset, sustainability will become increasingly meaningful in healthcare. And yet, in today's healthcare environment, leadership is just beginning to appreciate the benefits of sustainability. Why are healthcare leaders starting to care? It is because most leaders are worried about the economic future of their organization.

Few leaders cared about sustainability or energy efficiency when their revenue increased by 4 to 5 percent year after year during the 1990s. Leaders could apply increased dollars toward energy costs, facility expenses and labor inflation, and still invest 3 percent more in other costs of production, especially since many healthcare organizations were experiencing annual volume increases as demand expanded as well. In short, the economic model was stable and favorable for most healthcare providers.

SANDY SMITH: We've hardly scratched the surface of energy management. I don't even think we realize how much opportunity we have if we make energy management a priority.

FRED: You're right. I think we're headed in the right direction, though.

SANDY SMITH: You know, Fred, we received a letter from the utility company that was going to the state commission. We were named on the list of top energy users for our brand or public relations. Not so good for us to be on that list.

FRED: Agreed. I certainly care about how Memorial Hospital is perceived by the community. If (lack of) energy management is tarnishing our brand in a material way, then it becomes something we have to pay attention to. I think our challenge is difficult. I understand that much of the energy efficiency of our facility is built into the building when it's designed. So if you build it inefficiently in the first place, you have limited options later on. But I think the scoring mechanism and the reporting concept are interesting and probably useful for drawing more attention to energy efficiency as an issue. Do we have a sustainability section on our website? I thought that might be a good idea.

SANDY SMITH: Yes, yes we do.

FRED: Well, I mean, aimed at our patients and our customers where you [...]

SANDY SMITH: Yes, we do. We have videos up there for people to see what they can do to help Memorial Hospital be sustainable. And we have links to additional information, in case the public is interested. One of the things I would like to do on a regular basis is share our actual numbers; for instance, our energy savings. I think that doing so will be a good effort in terms of transparency and interdepartmental alignment.

FRED: Well, Sandy, as usual you're way ahead of me.

Opportunity cost

There is alignment between cost strategy and sustainability strategy as they pertain to effective energy use. If you, the leader, care only about driving down the cost of energy utilization, you would likely still be concerned about sustainability. Healthcare leaders want high quality at low cost. High waste means low quality and high cost, so healthcare leaders are concerned about all kinds of waste, including wasted energy. From a sustainability point of view, reducing waste is a means to an end; from a cost reduction point of view, sustainability is an end in itself.

Certain outcomes that serve a sustainability purpose are not necessarily going to serve cost, safety and quality outcomes. Such sustainability goals still require resources, whether that means time, attention or funds. What is the rationale for these "sustainability-only" goals? The rationale is founded on opportunity cost.

Dr. Hedge and Sandy Smith go their separate ways. Sandy Smith attends a meeting later in the day where Fred is present. At the end of the meeting, Sandy Smith connects with Fred and follows up on the sustainability conversation with Dr. Hedge.

FRED: Hi Sandy. Good to see you.

SANDY SMITH: Nice to see you too. You know, Fred, I've been giving more thought to our earlier conversation about sustainability. There's one question that I want to make sure I understand. I've been wondering whether there will come a time when sustainability isn't cost-neutral, as it often is today. If sustainability became a cost center that wasn't saving us money in operational costs on an annual basis, would sustainability be a program that makes sense to maintain?

FRED: There are programs today, here and now, that are not about making money. It's about providing a service to the community that inherently makes the big picture better. For instance, our pediatric services are necessary, but lose money year after year. It's not enough to bankrupt the system, but it's central to serving the community's needs.

SANDY SMITH: Some benefits to the community are high-value propositions, even though we lose money. So is sustainability any different from the pediatric programs you describe? Do the social and environmental benefits outweigh the costs to make it a high-value proposition for the community?

FRED: I think that, as a not-for-profit entity, it's clearly our responsibility to invest in those social and environmental components in a fiscally responsible manner.

It's a bit more challenging if you're an investor-owned or commercial enterprise. But I think it's a no-brainer for all organizations with a mission of health, and especially one that includes community service.

SANDY SMITH: I think you're on to something.

FRED: You know, Sandy, you manage to instigate change without making anybody mad. I mean, you're very artful at making stuff happen even when most people don't want to go along. Take that as a compliment. You know, leaders get people to do what they want them to do, and be happy about it. You have that leadership skill. I like this topic of energy management a lot. I can see so many ways that it applies to leadership, now that we've had a few conversations about it.

SUSTAINABILITY IN ACTION

Mark Platt serves as a member of the Board of Directors at S&S Cycle, Inc. He was previously a member of the Board of Trustees at Gundersen Lutheran Medical Center, and was a trustee at Gundersen Health System in La Crosse, Wisconsin. He is currently the Senior Vice President of Business Services at Gundersen Health System. Gundersen was the first hospital system in the nation to achieve energy independence by generating more energy with renewables (solar energy, wind energy and biomass) than it consumed. Gundersen experienced almost 80 days of energy independence in 2016, and continues to build its renewable energy portfolio.

What inspired the Gundersen Health System to become a leader in energy independence?

The former CEO, Jeff Thompson, was the sustainability visionary for Gundersen. The vision started with energy. Gundersen learned about energy intensity in the healthcare industry. The by-products the healthcare industry produces, like pharmaceutical and hospital supplies waste, worked against our mission to improve the health and well-being of our communities.

We live in a coal-heavy area, which worked against us. We did an energy audit in 2007 and realized we had a lot of work to do from an energy efficiency standpoint. We set a goal in 2008 to be energy independent by 2014. This goal was the catalyst to get us on a path for conservation. What started with conservation grew into energy independence. We looked at our regulatory environment and the resources available to us (wind, etc.). We started there. In October 2014, we had our first day as energy independent.

Why did reaching energy independence become your aim?

It was a matter of performing our mission with integrity. We believed that we could not only improve the health of the community by reducing our consumption, but we could improve the local environment, and do it all while reducing the cost of healthcare. This was a critical mission for us.

What does it take for a health system to become energy independent?

It takes a strong commitment. For us, becoming energy independent fits our culture and mission. Energy independence wasn't about headlines or saving a few bucks; it was all about being true to our mission. We invested $2.5 million into conservation efforts in our facilities. Now, we see $1 million return annually on these efforts. We had a brave, forward-thinking Board, and thought we should invest part of our savings into these conservation efforts. When we took 5 percent of our savings, it returned more than just dollars. We got back money as well as mission-critical health benefits. We wanted green dollars, not just green projects. We invested in a staff of engineers to help run energy projects. We hired these engineers from the local area. These engineers were Six Sigma Black Belt certified and perfectly qualified to help implement our energy projects.

How are Gundersen hospitals and facilities measuring their energy?

We're investing in metering. Our main hospital in La Crosse, four regional centers and clinics are all metered, which allows us to get data in real time. Our energy manager monitors these data. We look at energy bills and our meters to continually and closely measure our energy. We send out emails to employees when our meters indicate that there are high demand days, then direct teams to take action in response. Actions may include shutting off lights or reducing water consumption. High electricity demand days are typically in the summer when temperatures and air-conditioning loads are high. High electricity demand days are also on weekdays, when most team members are on-site.

What have been your biggest lessons learned along the way?

We learned a lot about energy generation and complex engineering projects. It's important to establish and maintain relationships with local power utility companies. Complexity and skillset around compliance is a high hurdle. Hiring engineers to go through this learning curve to build and implement energy projects has been very helpful. We are now exploring more solar projects.

On the conservation side, maintenance of the projects you pursue today is very important. Every year, we invest seven figures to reach forward in conservation and to maintain the projects we supported in previous years. All of this investment has had a positive return for us, both in terms of cost and other factors.

How did you convince your board members that Gundersen's mission to become energy independent is worth the commitment?

It started with a Board that was already laser focused on our mission. Mission was front and center. Our mission wasn't to maximize profit unless it improved the health and well-being of the communities we serve. Our CEO spearheaded this challenge; he wanted to reduce the cost of healthcare and improve the health of the community at the same time. This focus led to improved health, improved environment and improved financials. From our savings, we decided to fund energy projects. The money from savings would otherwise have been spent to invest in stocks, bonds and other initiatives. We decided to invest it in energy independence instead, which aligned with our mission.

What tools or resources have you found helpful in your healthcare sustainability journey?

We were the first in the industry to go down this path, so we had to identify our own tools and resources. We had to build the construction and engineering capabilities in-house. We had to build a team of engineers that believed in our mission and had the skillset to execute energy-efficiency projects. In short, we had a dedicated team. They had the technical capability. The resources we needed were the right people with the right values and skills to get these projects done. We needed a skillset that most clinicians and healthcare professionals don't have, and that's engineering.

Any advice for latecomers to sustainability?

It's easy to start! For people who haven't started, keep in mind that sustainability makes tremendous sense. For every dollar you invest in energy conservation, you could get 50 cents back annually. Typically, the return on investment is great, at about 30 to 100 percent each year. Make the investment and take the time to implement conservation efforts.

Food waste systems are also easy to implement. We invested US$100,000 into software to monitor food waste at Gundersen, and reduced that waste by 50 percent. This was an easy and quick win. You just need to know where to start.

Practical application: energy excellence at New York Presbyterian

The imperative

- Managing long-term risk from rising energy costs
- Achieving visibility as a leader within the industry
- Spurring innovation from employees
- Reducing costs
- Improving health by reducing air pollution.

The initiative

New York Presbyterian Hospital (NYPH) introduced its award-winning energy management program in 2003, and has since become an energy leader both inside and outside the healthcare industry.

NYPH started its energy journey by establishing a structure for managing the organization's energy initiatives. These efforts consisted of:

- Creating a full-time Energy Programs Manager position dedicated to maximizing the hospital's energy savings.
- Establishing the Office of Energy Management.
- Forming an energy team consisting of representatives from key departments.
- Adopting an energy policy.
- Becoming an Energy Star partner.

Once the energy management program structure was established, the organization's energy teams focused on identifying and pursuing key energy-saving opportunities. After performing several energy audits, NYPH implemented a variety of changes:

- It performed heating, ventilation and air conditioning (HVAC), controls system, lighting and central plant upgrades.
- With the help of a grant, it built a combined heat and power plant that achieved millions of dollars in annual energy cost savings and generated enough electricity to meet between 60 and 100 percent of electricity demands.
- It developed policies for identifying and exploring cost-effective purchasing solutions and energy-efficient practices for employees.
- It purchased a fleet of Energy Star-rated computers.
- It created engagement tools, such as a hotline number for calling in energy-saving tips and ideas, a reward program for employees who identify opportunities for energy efficiency improvements, a series of events to boost awareness with the public and staff, and a communications package for educating and sharing successes with employees.

These changes, along with many other energy conservation initiatives, helped NYPH achieve its status as a national energy leader.

The metrics

Energy performance is most easily benchmarked using tools such as Energy Star's Portfolio Manager, which is designed to track indicators such as:

- Quantity, type and cost of energy consumed (total and per square foot).
- Quantity, type and financial impacts of renewable energy produced on-site.
- Site energy intensity.
- Energy performance over time.
- Avoided greenhouse gas emissions from green power sources.

Lessons learned

- Forming a dedicated energy team is extremely helpful when working to develop a comprehensive energy management program. Capitalize on the organization's diverse intellectual resources by including experts from departments such as real estate, IT, engineering, facilities development, strategic sourcing and HR to develop the most comprehensive vision possible.
- Bring vendors, partners and other industry experts to the table for at least one energy-planning session to add diverse perspectives to the dialogue; vendors and partners are interested in helping the organization pursue its energy goals.
- Recognize that grassroots engagement is key to ongoing awareness and the long-term success of the energy program. Developing fun, interesting opportunities to get involved is crucial for achieving employee visibility and buy-in.
- Show commitment from the top to lend credibility and create hype around energy efficiency initiatives.
- Make the energy mission easy to understand and agree with, using familiar terminology to explain how the energy program helps achieve existing priorities such as cost containment, increased efficiency and long-term risk management.
- Identify energy performance improvement opportunities and prioritize investments in less efficient facilities by studying current and past energy. Performance may be gauged by comparing facilities performance with each other, to peers or to top industry performers.

About the organization

New York Presbyterian Hospital is the nation's largest nonprofit, nonsectarian hospital, encompassing 33 buildings and 8.6 million square feet. It is a leading academic medical center affiliated with two of the nation's leading medical colleges: Columbia University College of Physicians and Surgeons, and Weill Cornell Medical College. New York Presbyterian Hospital delivers comprehensive medical services to residents of New York City and its surrounding boroughs. It was the first healthcare system to earn the Energy Star Partner of the Year Award three times: twice for Excellence in Energy Management (2005 and 2006) and a third time in the Sustained Excellence category (2007).

Sources

Recycling Today Staff (2017). Study sheds light on millennials' stance on workplace sustainability. April 6. Available at www.recyclingtoday.com/article/rubbermaid-study-millennials-workplace-sustainability/ (accessed August 6, 2017).

www.energystar.gov/index.cfm?c=healthcare.bus_healthcare_ny_presb_hospital.

www.premiersafetyinstitute.org/wp-content/uploads/NYP-GreenCorner-0712 2010.pdf.

http://nyp.org/news/hospital/co-gen-plant.html.

Leadership imperative

When leadership embraces sustainability, there is a beneficial impact upon the health of the community, the environment and the planet. When sustainability was first introduced to healthcare in the 1990s, it was mostly a religious calling. People embraced sustainability because it was the right thing to do. Thus any discussion about the sustainability program costing money or fighting economic objectives was dismissed by fervent believers. The idea was that if you are into sustainability for the money, you are probably in it for the wrong reasons.

That mentality has changed. In today's landscape of the triple bottom line, it is encouraged and even expected to save money through sustainability practices that concurrently benefit the social and environmental dimensions of community health. In fact, anything short of these outcomes would not serve the triple bottom line. There is a sophistication to management that did not exist five, ten, 20 years ago. That newfound sophistication helps explain the high level of activity around exploring different service and operational models pertaining to sustainability and otherwise. For instance, many hospitals are outsourcing operations. They are migrating to a contractual model with pure-play businesses that expertly deliver higher quality, same or lower cost services than what the hospital could otherwise provide. While some aspect of sustainability can be outsourced, commitment to sustainability as a leadership imperative must be internalized.

Leadership dialogue

1   Sustainability is often an area of a leader's "conscious incompetence" when it comes to modeling leadership behaviors that are environmentally friendly. Why should leaders learn about the behaviors related to energy conservation and make a point of practicing such behaviors in private as well as publicly?
2   How should leaders navigate members of the leadership team who don't support efforts to reduce energy consumption and the hospital's carbon footprint, arguing that "energy costs savings are not material enough to bother pursuing an energy program"?
3   Measuring tangible indicators of energy consumption is a means of achieving team cohesion when metrics are made visible for all to see. What are example strategies that leaders can use to make energy metrics transparent and visible to both leaders and employees?
4   Sustainability-related solutions in energy management are sometimes "out-of-the-box" options that require transformational change. How can leaders encourage transformational thinking when employing team efforts to implement energy-impacting innovations?
5   The triple bottom line consists of fiscal, social and environmental responsibilities. Which bottom line is most often neglected by healthcare leaders while pursuing efforts to change organizational energy practices? Why?

Syllabus integration

Overview

Typical capital budgeting evaluations do not include environmental costs or savings, resulting in decisions that may prematurely dismiss energy management projects, despite potentially significant long-term environmental and financial savings. The capital budgeting evaluations described here will be applied to a variety of energy management approaches.

Objectives

1   To raise the economic valuation of energy management projects to a level that is worth considering in addition to alternative options.
2   To discuss key components needed to ensure the use of an energy management approach and how this approach can be institutionalized.
3   To utilize tools to measure, calculate and assess organizational energy usage:

   a   Quantify and allocate energy resources across appropriate business functions and activities.

4   To understand the link between human health and energy use.

Topics

1   Guidelines for energy management:

   a   Energy Star
   b   Energy Impact Calculator
   c   Energy savings.

2   Senate Energy Bill.

Homework

Environmental Impacts of the U.S. Health Care System and Effects on Public Health. Available at www.ncbi.nlm.nih.gov/pmc/articles/PMC4900601/.
2010 Healthcare Sector Energy Efficiency Indicator. Institute for Building Efficiency, Johnson Controls. Available at www.buildingefficiencyinitiative.org/articles/2010-healthcare-sector-energy-efficiency-indicator.
Develop a chart that highlights the health and financial impacts of energy usage from the US healthcare sector.

Additional resources

Energy Star (2016). *Healthcare Energy Savings Finacial Calculator.* Available at www.energystar.gov/buildings/tools-and-resources/energy-star-healthcare-energy-savings-financial-analysis-calculators.

GGHC (2007). *A Prescriptive Path to Energy Efficiency Improvements for Hospitals.* Austin, Texas: Green Guide for Health Care. Available at www.gghc.org/documents/Reports/GGHC_PrescriptivePath.pdf.

Journal of Healthcare Engineering (2013). Volume 5, No 2. *Energy Performance of Medium-sized Healthcare Buildings in Victoria, Australia – A Case Study.* Available at http://downloads.hindawi.com/journals/jhe/2014/974863.pdf.

Practice Greenhealth (2010). Greening operations series: purchasing tools and strategies to reduce energy and maintenance costs. Webinar, January 15. Available at www.practicegreenhealth.org.

The Energy Audit (n.d.). *Energy and Efficiency Case Study – Italian Mid-size Hospital.* Available at www.the-energyaudit.it/pdfProdotti/Case_Study_HealthCare.pdf (accessed July 30, 2017).

US Department of Energy (2011). Available at http://apps1.eere.energy.gov/buildings/publications/pdfs/alliances/hea_plugloads.pdf (accessed October 22, 2012).

Notes

1 Practice Greenhealth (n.d.). How health care uses energy. Available at https://practicegreenhealth.org/topics/energy-water-and-climate/energy.
2 Newby-Clark, I. (2009). Creatures of habit. *Psychology Today*, July 17. Available at www.psychologytoday.com/blog/creatures-habit/200907/we-are-creatures-habit.
3 US EPA (2004). *Leaders in Healthcare Tap the Power of Superior Energy Management.* Washington, DC: Environmental Protection Agency. Available at www.energystar.gov/ia/business/healthcare/factsheet_0804.pdf?9b81-81ae.
4 WHO and Health Care Without Harm (2009). *Healthy Hospitals – Healthy Planet – Healthy People.* Geneva: World Health Organization. Available at www.who.int/globalchange/publications/climatefootprint_report.pdf.
5 Energy Star (n.d.). About Energy Star. US Environmental Protection Agency and US Department of Energy. Available at www.energystar.gov/index.cfm?c=about.ab_index.
6 Hospitals and Healthcare Networks (n.d.). Containing energy expenses. Available at www.hhnmag.com/articles/5790-containing-energy-expenses.
7 Energy Star (n.d.). Building contest. Available at www.energystar.gov/index.cfm?c=healthcare.bus_healthcare_ny_presb_hospital.
8 Odell, A.M. (2007). Working for the Earth: green companies and green jobs attract employees. *GreenBiz*, October 16. Available at www.greenbiz.com/news/2007/10/16/working-earth-green-companies-and-green-jobs-attract-employees.

8

BEFORE A BABY'S FIRST BREATH

Safety + chemical management

FIGURE 8.1 Safer chemicals in healthcare

Executive summary

Healthcare organizations are high-stress, high-stakes environments in which safety is of primary concern to both patients and employees. Building a safety culture is part of building a sustainability culture. There are numerous examples of unsafe healthcare products that contain chemicals known to have negative health impacts. Kaiser Permanente developed a Sustainability Scorecard to vet the safety of such products, and made it freely available to the public.[1] Equally concerning are the chemicals used in healthcare whose interactive effects have yet to be studied. Increased awareness of environmental safety hazards places additional scrutiny on healthcare leaders to assure that healing environments are safe. Beyond direct patient impact, there are safety concerns related to the use and disposal of toxic materials.[2] This chapter helps readers understand when and how to use sustainability as a tool to mitigate safety risks related to potentially dangerous chemicals.

Storyline

Fred reads an article in Modern Healthcare *that he tears out of the magazine. The article features a story about a nursing group that questions adequate protection from chemicals in the workplace. Fred thinks: I need to share this with Kadine [Chief Nursing Executive] and Lori Prince [Vice President of Human Resources]. The following day, Fred has an opportunity to do just that. Kadine and Lori read the article and respond unanimously: "We follow all the applicable guidelines and safety protocols." Fred scratches his head, puzzled.*

FRED: But we don't have any downstream or follow-up studies. In other words, we don't have any data to know how our nurses have done *vis-à-vis* chemical exposure. I understand that these exposures are rare, but they could have a devastating impact if our nurses end up with cancer related to exposure.[3] This isn't just about meeting regulations. It's a concern that our employees may be exposed to harmful chemical agents. The safety of our people goes beyond regulation. [*Fred, not getting the reaction he is looking for, tells his assistant he'll be "right back" and goes off to visit with Lorenzo, Director of Environmental Services.*]

There is nothing more important to patients than safety

LORENZO LOPEZ: [*Director of Environmental Services*] Fred, what made you be concerned enough to tear out that article from *Modern Healthcare* magazine, or even pause to read the article? I know you must get bombarded with items to read.

FRED: I generally scan articles that come my way and express concern if I think most people would be concerned about a given topic or issue. My first job was running a big oncology clinic and, at the time, the nurses – nearly all women – had to gown and double-glove to give some of our chemo patients their medicines. And they found later that double-gloving didn't do the job it was supposed to do. So they were exposed to toxic substances. This was a few years ago, and now I find myself asking the same question: are we adequately protecting our employees? Here's the deal: safety is really a big issue for me.

BEFORE A BABY'S FIRST BREATH

Human beings are exposed to hundreds of chemicals every day. Chemicals exist in the foods we consume, in the clothes we wear and in the pillows we sleep on.[4] Does the average person pause to wonder what they are exposed to, and what the long-term impacts of the exposures are? What chemicals are bio-accumulating in human beings, and what chemicals are passed down from mother to child? The latter question was the subject of a study called *The Body Burden* conducted by the Environmental Work Group. In this study, the core blood of ten minority newborns was studied for the presence of chemicals, including bisphenol-A (BPA). BPA has become known for linkages to disorders such as cancer and obesity, and is also considered an endocrine disrupter.[5] In addition to the presence of BPA, 232 chemicals were found in the core blood of these newborns.[6]

"Please don't kill me"

The effective healthcare leader will increasingly be seen as a champion of patient and employee safety. At an elemental level, safety is the most important value to everyone. Freedom from iatrogenic harm is a core priority for patients. The equivalent priority for employees, including nurses and physicians, is personal safety – the notion of a safe and healthful work environment. Only after having a backdrop of patient and employee safety can employees focus on patient care.

Safety of the community: low incidence, high impact

Think about the worst things Memorial Hospital could do. The clinical team could operate on the wrong leg, or the wrong person, or leave a sponge in the body after surgery – no patient or community wants to hear that story. But safety translates beyond individual patients and employees. The community also tends to be upset when red bag waste is improperly disposed of in the regular landfill, because that waste has implications for the entire community. That is the worst fear at some level – in a very real way, there is a collective sense of "don't poison us." Chemical management entails asking, "What does exposure to chemicals mean to my health, the health of our employees and of our community?"

LORENZO LOPEZ: The link between chemicals and disease is becoming increasingly clear because of the increase in cancer and asthma rates.[7] Fred, are you aware of the potential impacts of the chemicals used at Memorial Hospital?

FRED: To be honest, Lorenzo, that's a level of detail that the average executive leaves to other people. Now, if I was directly responsible for employee health, if I'm Kadine, the CNE, if I'm the occupational safety compliance officer with OSHA [*the Occupational Safety and Health Association*] [...] if I'm in your position as the environmental health leader, then the answer would probably be "yes," because it is your business. Your job is built around those concerns.

LORENZO: [*Thinking to himself*] I think employee and patient safety are shared responsibilities of everyone on the hospital team.

MAKING PEOPLE SICK WHILE MAKING PEOPLE HEALTHY

Healthcare is often referred to as a calling rather than as a job. Pursuing this calling may require risky behaviors at times. It's a reality of daily work that healthcare professionals often deal with drugs and chemicals in treating the illnesses of their patients. The impact of the exposure to these chemicals, such as chemotherapeutic drugs, by caregivers is being called into question.[8] A study produced by Physicians for Social Responsibility investigated the exposure of healthcare workers to chemicals. The results highlighted that, among other chemicals of concern, bisphenol A, phthalates, polybrominated diphenyl ethers (PBDEs) and perfluorochemicals (PCFs) were found in all individuals tested.[9]

Think about where you're going, rather than where you are today

Fred rereads the article from Modern Healthcare *magazine about a nursing group that questions adequate protection from chemicals in the workplace.*

FRED: I'm trying to deal with what I think might be coming rather than where we are today. Is chemical management evolving into something that we ought to be more aggressive about tracking? Is the evidence suggesting that chemical management is a more serious danger to our nurses than previously imagined? Those are the questions I am trying to explore. And the answer I get from our leaders was, "We're complying with all of the current rules and regulations. We're doing everything that the law requires" [*sigh*]. Well that's great, but if we're doing what we're required to do, but are trading on the health of our employees, shouldn't we do more than what the law requires? I want to know why and how we're going to impact our employees in the future, not just today. Once upon a time, people were smoking in the staff lounges. Now we prohibit smoking on the entire campus. Why did we make this a smoke-free campus? We – along with other healthcare organizations – prohibited smoking before any law to prohibit smoking existed, because it was the right thing to do.

LORENZO LOPEZ: Was that the only reason why we prohibited smoking on campus? Today, there's evidence demonstrating that smoking causes cancer and kills people and that second-hand smoking is bad for other people, not just the smoker. Did we prohibit smoking on the campus because of all the mounting evidence that smoking is unsafe and unhealthy? Maybe we should approach chemical management the same way.

FRED: Yes, there was mounting evidence that smoking was harmful to health, but that had nothing to do with the decision to eliminate smoking on campus.

LORENZO LOPEZ: Well then, why did we make that change?

FRED: You know, Lorenzo, the most persuasive argument for eliminating smoking had everything to do with the other healthcare campuses that migrated to a no-smoking policy. That was the most persuasive argument: everyone else is already doing it and various professional organizations recommended it. If we didn't have groups like the American College of Physicians recommending smoke-free campuses, very few organizations would follow through, because few leaders and physicians are going to individually make the decision to change. That's the same reason why there's a tipping point around sustainability (Gladwell, 2002). Once organizations across the country adopt chemical management as a safety imperative, leaders will say, "The other leaders and physicians are doing it, the accrediting bodies recommend it, the nurses are for it, the Joint Commission thinks it's a good idea. We better adopt the new standard." But until the chemical management issue makes its way into professional associations and accreditation standards, culture change is unlikely.

LORENZO LOPEZ: That all makes sense, Fred. But there are people on the frontline who are early adopters. For instance, Dr. Bea and Dr. Cohen came to me the other day asking why we're not managing our chemicals more proactively and what they can do to kick-start the process.

CHEMICALS OF CONCERN IN HEALTHCARE

The following is a list of 11 chemical types that cause asthma:

- Cleaners, disinfectants, sterilants
- Natural rubber latex
- Pesticides
- Volatile organic compounds
- Baking flour
- Acrylics
- Fragrances
- Phthalates
- Environmental tobacco smoke
- Biologic allergens
- Drugs.

Sacred trust

Physicians have a "sacred trust": the notion that they have been given talent and training to care for people at their most vulnerable stages of life. With that sacred trust comes a tremendous power and personal accountability, because patients put their lives in clinical hands. This is where most clinicians define the limit of their personal accountability in terms of accountability to the individual patient. Issues of equity or group fairness are often perceived to be beyond the traditional scope of a clinician's responsibility. This reality explains why focusing on safety as a leadership priority is necessary for the health and well-being of society.

One of the realities of modern healthcare is that there has been, and still is, scientific support for only a fraction of the treatments and practices employed today. Data are what inform evidence based medicine, but there are simply no statistically significant data for every care process. Clinicians take patients' lives into their hands knowing full well that this is the reality of the situation. That is why medicine continues to be a wonderful mixture of art and science. It breeds a cautious conservatism that is foundational to medical practice.

The relative lack of iterative processes in medicine explains the relative lack of change in clinical processes used in caring for patients. Some clinicians may not agree with that statement, though many would agree that there is not much room for error in medicine. Physicians would say that they learn from experience and personalize care by making tweaks to the processes they were taught in school. In truth, it is hard to see much change in the fundamental approach to medical practice as compared with 50 years ago.

FRED: I know that, as the leader, I'm responsible at some level for that lack of progress.

LORENZO LOPEZ: I can be slow, but still faster than 90 percent of the crowd I hang out with. The "cream of the crap," as I like to say.

FRED: I was speaking with a friend of mine from graduate school who is leading a healthcare organization in the Midwest. She told me an amusing story. She was sitting in on a physician meeting. This young physician appealed to his peer group of pediatricians, trying to convince them that toxic chemicals can be a tremendous hazard to children. He said physicians had a responsibility to ensure that parents prepare their home environments to be toxin-free to the extent possible, prior to taking home a newborn. That could be a new business line, now that I think about it. [...]

TOXIC SUBSTANCE CONTROL ACT

The US EPA has been pushing for reform of the Toxic Substance Control Act. Current regulations do not require disclosure of chemicals used in many healthcare products.[10] In addition, the EPA does not have the authority to establish safety standards related to chemical management.[11]

Professional associations set the default for practice change in medicine.[12] When in doubt, clinicians typically reach out to their professional associations to determine the ethical and/or practical way(s) forward. Otherwise, most clinicians are incentivized to act on individual decision-making authority, driven by doing what is best for the patient, in their judgment.

Let us look at how the Joint Commission helps determine healthcare culture. For many years, the Joint Commission published an extensive book of requirements. Periodically, the Joint Commission evaluates a given hospital's performance in terms of patient safety. There is a set of prescriptive procedural requirements, and the task is to fit each one of those procedural requirements into operations to create an environment that prioritizes safety and quality. This system promotes a culture of working to standard and not one of continuous improvement. Working to standard does not incorporate the concept of innovation, but rather the simple discipline of faithful repetition.

Leaders should prioritize

Engagement – the notion of everyone being connected to the change – is a key driver of culture change. Engagement is the opposite of the traditional command-and-control mentality that defines much of the current healthcare culture. Engagement is about being team-oriented; it is about empowerment for innovation and change; it is about trying new things. Engagement is not only about institutionalizing sustainability practices, but also about how we design healthcare work processes and ask questions. Look at General Electric or Apple, two successful companies known as innovative businesses that deliver exquisite products. The values of teamwork, innovation and empowerment are defining values of both companies.

It is this type of culture that empowers employees and physicians to ask, "What is the best way to keep people safe?" Perhaps chemical management is part of the answer. Regardless of where chemical management falls in the leadership priority list, safety as a leadership priority is certainly at or close to the top.

Promoting an iterative learning culture should be on the leader's priority list as well. Excellent leaders are not afraid of change and, in fact, aggressively promote change that moves the organization into alignment with its customers' priorities.[13] Leaders empower teams to create a high rate of change that is accretive to customers. When organizations have physicians in leadership roles, they are often culturally sophisticated enough to lead change more rapidly within physician populations.

In short, the professionalization of leadership drives the best leaders to be the best change agents, whether in respect to toxic chemical management, or clinical process improvement, or patient safety. Upon taking a closer look, leaders do not make changes. Rather, they engage and empower others to be change agents.

FRED: The people in Occupational Health, for instance, know we ought to be worried about chemical management. They know the challenges that are on the horizon. They know the chemicals that are really bad. My job as the leader is to communicate with our Occupational Health professionals and infuse them with a cultural bias that causes them to say, "You know, we ought to get rid of these harmful chemicals."

LORENZO LOPEZ: How do I empower people to make chemical management a safety priority?

FRED: Let's say the evidence mounts that toxic chemicals are bad. They're bad for our patients, bad for employees and bad for the community if we don't dispose of them properly. The evidence of these realities mounts to the point that employees come forward to you and to Occupational Health recognizing that chemical management is an issue. How do you then empower that message to be delivered and changes to be made?

LORENZO LOPEZ: Well, um […]

FRED: Let's go back to where we started this conversation, with the article from *Modern Healthcare* magazine about the harmful chemical agents. I was looking to the future of where this issue is going, rather than where we are right now. Sharing that article about chemical and nurse safety sent a message that I thought it was important. I didn't empower our colleagues to do anything about it, though. I didn't make it clear enough to Kadine [*Chief Nursing Executive*] or Lori [*Vice President of Human Resources*] that chemical management is an organizational priority.

MANAGEMENT TIP

One key job of leadership is to change the culture so that changes percolate up from the people closest to the problem, rather than depending on top-down policy pronouncements. Creating a culture where people close to the problem feel empowered requires the articulation of priorities, with safety being high on the list. Once the safety priority is articulated, the next step is to empower employees to engage the safety value within their sphere of influence.

Effective leaders create a culture of accountability where employees and physicians feel accountable to make changes happen within their sphere of influence. With a culture of empowerment and accountability, employees will promote sustainability, just like safety, if sustainability is a community priority. To date, healthcare chemical management issues and concerns are not sufficiently defined to evoke a regulatory response.[14] Those standards that do exist are largely outdated and lack environmental perspective.

SUSTAINABILITY IN ACTION

Kathy Gerwig is Vice President, Employee Safety, Health and Wellness, and Environmental Stewardship Officer for Kaiser Permanente, one of America's leading not-for-profit health systems. Kathy is responsible for developing, organizing and managing a nationwide environmental initiative for the organization. Under her leadership, Kaiser Permanente has become widely recognized as an environmental leader. Her book, *Greening Health Care: How Hospitals Can Heal the Planet*, examines the intersection between healthcare and environmental stewardship. Kathy is also on the boards of several leading non-governmental organizations focused on safety and environmental sustainability in healthcare.

What led Kaiser Permanente on the journey to become more sustainable?

We realized that there was a clear connection between negative environmental conditions and disease. Our model of healthcare focuses on prevention as well as treatment. Preventing the environmental causes of illness was the main objective of our sustainability program. We also recognized that the healthcare industry's impact on pollution and waste was substantial, and we wanted to be proactive in combatting it.

How did Kaiser Permanente go about beginning to implement its sustainability mission? How do those initial steps compare with where your sustainability program is today?

We started our efforts about 20 years ago, and the sustainability landscape was very different back then. Our initial priorities were to eliminate mercury from our operations, reduce the amount of waste we generated, and to obtain building materials and medical products without polyvinyl chloride (PVC). In addition to being good for the environment, these priorities supported the safety of our workers and patients. We could make a good business case for the work, which was a in from health, business and reputation standpoints. At the time, sustainability efforts seemed daunting, but successes have grown across the marketplace. For example, in the past two years, our decision to avoid chemical flame retardants and antimicrobial treatments in furnishings has been adopted by other hospitals' systems, shifting millions of dollars of purchasing to safer products.

How do these initial steps compare to efforts today? Sustainability initiatives have gotten much more sophisticated because the health impacts of climate change and pollution are at a crisis point. The healthcare sector's role in addressing the problems requires us to be much more visionary.

What challenges did you have to overcome as you were changing the industry and marketplace?

Two challenges come to mind. First, the marketplace wasn't ready for what we wanted, and is still not totally where we want it to be. We sometimes have a difficult time getting the products we want from our suppliers. For example, we want PVC-free products, but PVC is commonly used in medical products, building materials and office products. The marketplace hasn't been keeping pace with where we want to go. However, healthcare systems (Kaiser Permanente and Inova in particular) have signaled to the marketplace where we want to go, and manufacturers and suppliers are starting to respond. These innovators want to make the change to more sustainable products. This is fantastic to see.

Second, some industry associations are very powerful and well-funded. Manufacturers have a stake in maintaining the status quo. It can be difficult to go up against these stakeholders; it's a huge barrier.

The subtitle to your book is How Hospitals Can Heal the Planet. What advice do you have for other hospitals and industry players seeking to adopt green practices?

Hospitals need to look at every dime they are spending. Everyone is trying to reduce costs. Understanding the cost implications of any decision is important. We look at the total cost of ownership. Almost every environmental/ sustainability decision we have made at Kaiser Permanente has been at a cost saving. Chief Financial Officers need to understand the total cost of energy based on different sources, including wind and solar. When looking at greener products, there are certifications on some product lines that are very helpful. We rely on some of those certifications to identify green products, and that helps us determine the total cost of ownership of our purchasing decisions.

There also needs to be learning and collaboration. There are so many resources out there. This is a space where hospitals are working with each other and with NGOs to make sustainability effective. Employees need to be engaged. Green teams can be created to seek out advice from employees to come up with solutions. There are certain activities that need to happen at a unit/individual level (e.g., waste management, printing, recycling) that require the engagement of employees. For example, we have employees on water watch in California, where recent drought conditions made conservation an even more pressing issue. We need leadership support, system support *and* employee engagement.

What do you see as the future of sustainability in healthcare?

I feel the future is centered on community health in a broad sense. There are requirements for nonprofit hospitals to assess community needs and prepare plans to address them. We expand that work to include environmental actions that also support those needs. If a community has obesity problems, for example, a hospital system can help address them through improvements that promote walking and bicycling and the use of public transit. It can develop programs that get people exercising, while reducing energy consumption and air pollution.

We can take a broad view and think about the role of the hospital as an anchor in the community. Kaiser Permanente wants to collaborate with local organizations to support purchasing from local suppliers, promote healthy watersheds and create good jobs. Local partnerships help us leverage the special roles hospitals have in communities.

What tools or resources have you found helpful in your journey in sustainability healthcare?

The main resources we use are Health Care Without Harm and its Practice Greenhealth subsidiary. Our journey would have been less successful without these partners. Having shared metrics with other hospital systems has also been big. Some hospitals have agreed to measure and report the same sustainability actions in the same way, which is important to help mature the sustainability contributions hospitals can make.

Practical application: Nordic Swan for Healthcare targets polyvinyl chloride/Di(2-ethylhexyl)phthalate

The imperative

- Driving the market for alternatives to products containing chemicals of concern.
- Reducing the toxicity of healthcare waste.
- Minimizing risk to human health.
- Fulfilling the organizational mission to protect public health.

The initiative

In the global market, the Nordic Ecolabelling organization has developed one of a few "eco-standards" that target chemicals of concern in medical products.

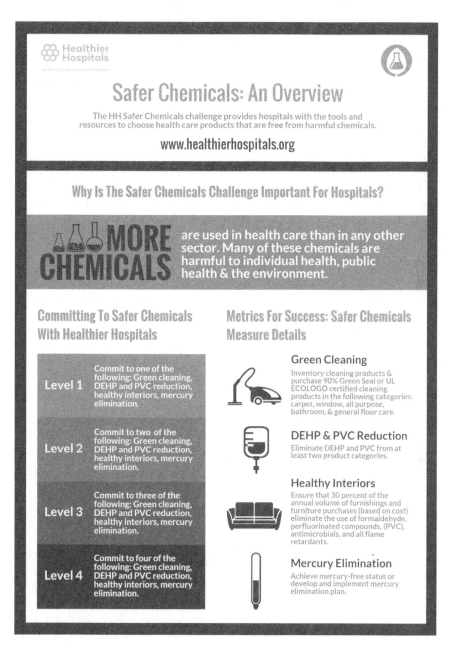

FIGURE 8.2 Safer chemicals: an overview

The primary objective of this label is to help healthcare organizations in the Nordic region identify products that reduce chemical exposures at every stage of the product life cycle.

Manufacturers and distributors of disposable bags, tubes and accessories for healthcare who mark their products with the Nordic Swan Ecolabel commit to eliminating polyvinyl chloride (PVC), harmful plasticizers and a variety of other chemicals that are harmful to health. They also commit to reducing packaging waste and ensuring consistent levels of safety, quality and performance in their products.

Although standards such as these do not yet exist in most countries – including the US – many other resources exist for targeting chemicals of concern in health-care products. Sample questionnaires may be used to help spur a dialogue with vendors; sample safer chemicals or environmentally preferable purchasing (EPP) policies can help specify purchasing priorities; case studies can help provide insights into the key success factors in implementing a safer chemicals program.

To help guide these processes, experts in chemical safety in healthcare identified a priority list of chemicals to eliminate based on environmental and health impacts:

- BBP
- DBP
- DINP
- DNOP
- Mercury
- Persistent bioaccumulative toxics
- Lead
- Potentially endocrine-disrupting plasticizer phthalates (DEHP, n-butyl benzyl phthalate (BBP), Dibutyl phthalate (DBP), Diisononyl-phthalate (DINP), Di-n-octylphthalate (DNOP))
- Halogenated plastics, including PVC
- Flame retardants (halogenated, chlorinated, brominated)
- Carcinogens, mutagens and reproductive toxics
- Bisphenol A (BPA)
- Volatile organic compounds
- Latex.

The metrics

Measuring the progress of a safer chemicals program can be achieved by tracking:

- Qualitative metrics: presence of a safer chemicals policy, existence of a value-add in purchasing process for products with reduced chemicals.
- Quantitative metrics: total dollars spent on safer chemicals products, percentage of safer chemicals products purchased (total, within a product category, within a department), total number of chemicals of concern targeted.

Lessons learned

- Nordic Ecolabelling has stated that it will consider including requirements regarding energy, recyclability of materials, labeling and design, process contaminants and residues, and working environments in future criteria, alongside its current criteria focusing on safer chemicals. This alignment highlights the close connection between developing a safer chemicals program and an Environmentally Preferable Purchasing program – the two priorities are closely connected and can be built together into policies, purchasing practices and conversations with suppliers.

• Several groups within and external to the healthcare industry are working to identify chemicals of concern in hospitals and the products that contain them. Examples of such groups include the United States Department of Labor, Physicians for Social Responsibility, and Kaiser Permanente and Health Care Without Harm. Targeting the areas where work has already been done is the easiest place to start.

• For organizations that are unsure where or how to start implementing a safer chemicals program, there are several easy-to-use tools to help healthcare organizations identify chemicals of concern, track the products they are in and develop a plan for eliminating them.

• Engaging the hospital's purchasing department in the planning process and educating them extensively about the safer chemicals program is crucial. Talk about the chemicals being targeted, including their health impacts and what products they are found in, so that the entire purchasing team is knowledgeable and capable of communicating about the sustainability program with vendors.

About the organization

The Nordic Ecolabel is the official Ecolabel of the Nordic countries, and was established in 1989 by the Nordic Council of Ministers with the purpose of providing an environmental labeling scheme that would contribute to sustainable consumption. Covering over 60 product groups, it is a voluntary, positive ecolabelling of products and services.

Sources

Guide to Choosing Safer Products and Chemicals. Available at https://noharm. org/sites/default/files/lib/downloads/chemicals/Guide_to_Safer_Chems.pdf.
Nordic Ecolabel for disposable bags, tubes and accessories for healthcare. Available at www.nordic-ecolabel.org/criteria/product-groups.

Leadership imperative

Safety should be on every healthcare leader's priority list. Similarly, sustainability should be on every healthcare professional's priority list. Sustainability improves individual and communal safety, and prevents harm in cases such as chemical management. It is the leader's responsibility to create a culture where employees and physicians know that chemical and environmental safety is an organizational priority.

In productive, safe organizational cultures, employees are encouraged to break down silos and work across boundaries to address issues of safety and sustainability.

In cultures that collaborate across departments, two things happen related to chemical management. First, employees and physicians feel accountable for identifying issues related to toxic chemicals or similar problems. Second, employees and physicians understand enough about leadership to socialize chemical management, thus bringing about change in the professional setting.

Many leaders reflexively turn to top-down solutions because they seek a quick, easy, short-term solution. They wait for the law or regulation to influence chemical management practices in healthcare long term, which tends to be the least efficient or effective method due to lack of employee engagement. The leadership learning embedded in this chapter is that chemical management is not on the leader's priority list, but safety certainly is. Leaders will turn to organizational experts for specific guidance about toxic chemicals. As evidence mounts that healthcare needs to improve chemical management processes, leaders will increasingly recognize that – like any other sustainability topic – chemical management is ultimately about safety, accountability and facilitating a culture of empowerment.

Leadership dialogue

1 What are possible sources of camaraderie and tension between leaders in healthcare and leaders in sustainability regarding environmental policies and procedures?
2 Hospital safety programs are largely concerned with patient, employee and visitor safety. These are the foci of advocacy for sustainability, as well as the safety of the community. In what ways do these safety concerns align and diverge?
3 Why should – or should not – leaders wait until substances are listed on the regulatory "proven dangerous" list before taking action?
4 How does a leader create employee and physician accountability for sustainability within a healthcare organization?

Syllabus integration

Overview

Review product evaluations, focusing specifically on their documentation of quality impacts as well as workforce safety and health. Students will familiarize themselves with the numerous examples of unsafe healthcare products that contain chemicals known to have negative health impacts. This class helps students understand when and how to use sustainability as a tool to mitigate safety risks related to potentially dangerous chemicals.

Objectives

1 Reconsider chemical management to incorporate sustainability and public health values.
2 Study how chemical management affects community relationships and business relationships with physicians, manufacturers and product suppliers.

3   Understand why safe chemical management supports prevention and wellness goals.

Topics

1   What to do in worst-case scenarios
2   Key safety concerns of chemical management
3   Green cleaning:

    a   Clinician perspective.

Homework

Chase Wilding, B., Curtis, K. and Welker-Hood, K. (2009). *Hazardous Chemicals in Healthcare: A Snapshot of Chemicals in Doctors and Nurses.* Washington, DC: Physicians for Social Responsibility.

EWG (n.d.). 232 toxic chemicals in 10 minority babies, Environmental Working Group. Available at http://articles.mercola.com/sites/articles/archive/2009/12/31/232-toxic-chemicals-found-in-10-babies.aspx.

Healthier Hospitals (2017). *Safer Chemicals.* Available at http://healthierhospitals.org/hhi-challenges/safer-chemicals.

WHO Europe (2014). *Health-sector Involvement in Chemical Management at a National Level.* Available at www.euro.who.int/__data/assets/pdf_file/0020/242660/Health-Chemical-Web_Final.pdf.

WHO (2017). *Chemicals and Health-care Waste.* Available at www.wpro.who.int/health_environment/topics/chemicals_and_wastes/en/.

Additional resources

Markkanen, P., Quinn, M., Galligan, C. and Bello, A. (2009). *Cleaning in Health-care Facilities.* Health Care Without Harm and the Global Health and Safety Initiative. Available at www.sustainableproduction.org/downloads/CleaninginHealthcareFacilities.pdf.

US EPA (n.d.). *Essential Principles for Reform of Chemicals Management Legislation.* Washington, DC: Environmental Protection Agency. Available at www.epa.gov/assessing-and-managing-chemicals-under-tsca/essential-principles-reform-chemicals-management-0.

Notes

1 Anon. (2010). Kaiser Permanente unveils sustainability scorecard for medical products. Environmental Leader: *Environmental and Energy Management News*, May 4. Available at www.environmentalleader.com/2010/05/04/kaiser-permanente-launches-sustainability-scorecard-for-medical-products/.

2 Dresser, T. (2012). Hazardous waste generation & management. US Environmental Protection Agency, May 22. Available at www.epa.gov/region7/education_resources/teachers/ehsstudy/ehs9.htm.

3 National Cancer Institute (n.d.). President's cancer panel – annual reports. National Cancer Institute. Available at http://deainfo.nci.nih.gov/ADVISORY/pcp/annual Reports/index.htm.

4 Highfield, R. (2006). The reality is that everything is made of chemicals. *Telegraph*, November 8. Available at www.telegraph.co.uk/news/uknews/1533519/The-reality-is-that-everything-is-made-of-chemicals.html.

5 Breast Cancer Fund (n.d.). Chemicals glossary: bisphenol A (BPA). Breast Cancer Fund. Available at www.bcpp.org/resource/bisphenol-a/.

6 EWG (n.d.). Toxic chemicals found in minority cord blood. Environmental Working Group. Available at www.ewg.org/news/news-releases/2009/12/02/toxic-chemicals-found-minority-cord-blood#.WZN5dNPyu8U.

7 Poladian, C. (2012). Asthma rates continue to soar in the US. *Medical Daily*, May 17. Available at www.medicaldaily.com/asthma-rates-continue-soar-us-240475.

8 Physicians For Social Responsibility (n.d.). Hazardous chemicals in health care. Available at www.psr.org/resources/hazardous-chemicals-in-health.html.

9 Ibid.

10 US EPA (2010). Essential principles for reform of chemicals management legislation. US Environmental Protection Agency. Available at www.epa.gov/assessing-and-managing-chemicals-under-tsca/essential-principles-reform-chemicals-management-0.

11 Ibid.

12 AMA (n.d.). AMA's Code of Medical Ethics. American Medical Association. Available at www.ama-assn.org/ama/pub/physician-resources/medical-ethics/code-medical-ethics. page.

13 Kanter, R.M. (2008). A financial turnaround requires culture change. *Harvard Business Review*, October 29. Available at http://blogs.hbr.org/kanter/2008/10/a-financial-turnaround-require.html.

14 US EPA (2010).

9

DOWNSTREAM WITHOUT A PADDLE

Quality + Environmentally Preferable Purchasing

FIGURE 9.1 Environmentally Preferable Purchasing: the front door to sustainability

Executive summary

The community's quality of health is, in part, dependent on keeping the environment free of toxic chemicals.[1] Waste management and chemical management both impact this subject. They are downstream mitigation methods once toxic substances have already been used and have the potential to create negative impacts. Environmentally Preferable Purchasing (EPP) is the front door to sustainability because it prevents toxic substances from entering healthcare facilities in the first place. The healthcare industry has the opportunity to act as an agent for change in environmentally sustainable market trends by employing its large purchasing power. Across the country and the world, hospitals are partnering with group purchasing organizations (GPOs) and manufacturers to eliminate toxic chemicals from healthcare products and facilities. This chapter is a resource for leaders to explore why and how EPP opportunities are a means of achieving quality goals.

A told story

Fred has a bi-weekly mentoring meeting with Felipe, Vice President of Materials Management at Memorial Hospital, during which he discusses how to successfully integrate core elements of sustainability into the hospital culture. During this conversation, Fred and Felipe discuss how to engage employees in understanding the impact of EPP upon their work.

FRED: Felipe, how is your day going?

FELIPE: Frustrating. My car's check engine light went on. I had to take it to the shop again. This is the third time in as many months. I'm starting to wonder about the quality of the car.

FRED: Huh, that's interesting. Imagine that instead of repairing your car after repeated incidents, the mechanic taught you how to prevent the check engine light from going on in the first place.

FELIPE: I wish that would happen, but that's not reality. The car mechanic told me, "Preventing the light from turning on is not my problem. My problem is that we need to fix the car quicker, faster and cheaper. We have a business to operate, you know."

FRED: I see. So the mechanic is mostly incentivized by the volume of cars he fixes. And now you're wondering about the quality of his work because you've visited so frequently. I use a mechanic who fixes the root problem. Now that's a quality mechanic!

FELIPE: I see what you're saying about targeting and fixing the root problem, but I have to admit, I am not sure what the mechanic has to do with anything […] other than perhaps I should find a better mechanic.

WHAT IS EPP?

Environmentally Preferable Purchasing is the procurement of products and services that, through their life cycle, do not cause harm or cause less harm to the environment, patients, employees or community at large.

SUSTAINABILITY IN ACTION

Beth Eckl is Director of the Environmental Purchasing Program for Practice Greenhealth. She provides technical support, consulting services and training on environmental purchasing for health systems, healthcare facilities and group purchasing organizations. Beth oversees the development of environmental criteria for dozens of medical and non-medical products used in hospitals. She is also responsible for the EPP Specifications and Resources Guide for Health Care, the first and only source for healthcare supply chain professionals.

How/what has the evolution been over the past ten years in terms of supply chain sustainability in healthcare?

Let's look further back than that. In the late 1990s there was a memorandum of understanding between US EPA, the American Hospital Association and Hospitals for a Healthy Environment (now Practice Greenhealth) to address pollution prevention in healthcare. The concern was that the healthcare industry was releasing significant amounts of mercury and dioxin into the environment and generating considerable waste – both would affect the health of people and communities that hospitals serve. One way to reduce the release of mercury into the environment was to eliminate mercury coming in through the supply chain and replace mercury-containing products with those that were free of mercury. Mercury and dioxin in waste was one of the first major sustainable supply chain issues in healthcare, gaining more momentum when five of the largest GPOs (group purchasing organizations) signed an agreement to purchase mercury-free products.

When I came to Practice Greenhealth in 2009, we were working with five GPOs as members. About 95 percent of hospitals purchase from one or more GPOs, so having environmentally preferable products available from GPOs was important. We provided technical expertise on every facet of sustainable procurement by prioritizing purchasing contracts, educating contracting staff and suggesting environmentally preferable purchasing criteria. We targeted contracts where mercury was a concern and provided information on available alternatives. Our sustainable purchasing priorities also evolved on other issues, including chemicals of concern in healthcare products, waste reduction, energy and water efficiency, and healthier food. Health Care Without Harm campaigned to eliminate the use of polyvinyl chloride (PVC) – as well as the plasticizer typically used in PVC, DEHP (di-2ethylhexyl phthalate) – in medical plastics, particularly in the NICU where the health impacts from exposure to DEHP were of greatest concern. Efforts were also underway to minimize waste by embracing reusable instead of disposable products. Many Practice Greenhealth members found that cost savings with the environmentally preferable alternatives demonstrated a clear business case. Kaiser Permanente also created one of the first Environmentally Preferable Purchasing positions in the healthcare industry.

One of the outcomes of exploring sustainability in supply chains was the need for product transparency. Most of the GPOs were asking suppliers detailed questions about specific attributes in bidding documents that had to be answered in a short window of time. To improve the quality of the information received and to educate suppliers on key criteria, the five largest GPOs agreed to standardize the environmental attributes asked. In 2011, we collaborated with the five largest GPOs to gain agreement on a set of standardized environmental questions to ask suppliers. The goal was to ensure that safer products were used in healthcare to reduce environmental and human health impacts, to improve the quality of information GPOs were receiving, and to educate suppliers and others on the important environmental criteria to healthcare. This initiative was also aligned and informed by Kaiser's sustainability scorecard.

In 2012, Practice Greenhealth launched the Healthier Hospital Initiative (HHI), a three-year campaign originally founded by 12 of the largest, most influential US health systems, comprising over 500 hospitals with more than $20 billion in purchasing power, along with Practice Greenhealth, Health Care Without Harm and the Center for Health Design. HHI was a guide for hospitals to reduce energy and waste, choose safer, less toxic products, and serve healthier foods. HHI provided more than 1,300 hospitals of all sizes with free tools and resources that help focus sustainability efforts on the healthcare sector's biggest areas of opportunity and risk. Our program participants were seeing amazing results – using less energy and fewer chemicals as well as increased healthy foods served. As a comprehensive, sector-wide initiative, our program was fully consistent with and builds upon the Centers for Medicare and Medicaid Services' (CMS) Triple Aim: "Better Health, Better Care and Lower Costs." It instituted frameworks and implementation tools that may be used by every healthcare facility in the nation.

Because of the momentum and success of the Initiative, Healthier Hospitals has become a permanent free program of Practice Greenhealth. HH continues to use data to engage around proven environmental strategies and lead communities to a healthier future.

How do you envision the healthcare sustainability industry will grow?

There has been growth in sustainable procurement and a few surveys attest to this growth.

A *survey*[2] of healthcare professionals found that 54 percent of hospitals say that green attributes are very important in their purchasing decisions, and 80 percent expect that to be the case in two years. The survey revealed the business and social benefits of sustainable products in protecting hospital staff (78%), improving health outcomes (55%) and making financial sense (60%).

Another survey of 120 global supply chain professionals found that almost all organizations within the surveyed group (97%) place a high level of importance on sustainable procurement. The report states that this trend has been growing in the healthcare procurement field globally.

As the link between environmental and human health becomes stronger, sustainability in healthcare and the recognition of solutions through procurement will continue to grow. In terms of how sustainable procurement will grow, look for more cost-effective, environmentally preferable products on the market which dispel the myth that all green products cost more. One method of increasing the availability of competitively priced sustainable products is to look at costs beyond the purchase price. As hospitals seek cost-saving solutions, assessing products based on the full costs to an organization could provide the business case. For example, when you buy a car, you typically look at the cost of fuel, insurance, repair, etc. – not just the purchase price – to know your total costs as a car owner. Likewise, a $99 copier may have ink cartridges that cost twice as much as a more expensive copier, so it is important to look at the total cost of ownership: price, usage costs (such as accessories, labor, energy, water) and end-of-use costs (such as waste). In the same way, hospitals which purchase products that reduce waste and save energy will see associated cost savings when they look beyond the purchase price. Practice Greenhealth has developed an easy-to-use *Cost of Ownership Calculator*, v1.0, to enable a cost assessment and drive cost-saving solutions for sustainable products.

Growth in sustainable procurement will occur in product areas ripe for market transformation and with a common standard. A few years ago, we launched two working groups of hospitals and health systems with the purpose of transforming markets around healthier food and safer chemicals. The power of these working groups has been the ability to bring together large, leading hospital buyers and work with them to create ambitious yet achievable environmental specifications that influence product design and production decisions by producers. With the Safer Chemicals group's focus on healthy interiors, for example, more than 35 major furniture suppliers have provided lists of products that meet our criteria for furnishings without flame retardants, PVC, perfluorinated chemicals, antimicrobials and formaldehyde. These criteria now reach well beyond the healthcare sector. Government, higher education and other institutional furniture purchasers are now using our furniture lists to identify products that will help them achieve their organizational sustainability goals.

As important, after two years of work with the Business and Institutional Furniture Manufacturers' Association (BIFMA), our criteria will soon be embedded in BIFMA's major furniture sustainability standard. Working with a vocal and influential group of health systems, this working group has moved the furniture sector to both remove the targeted chemicals of concern and also to actively market these offerings to a broad range of purchasers. Moreover, they were able to transform not only the institutional furniture sector but also the medical furnishings sector which is now providing products that were not on the market only a year ago.

What are the greatest levers for systematic change?

The healthcare sector itself is a lever for national change because it is already committed to health and healing. It can be an embarrassing and uncomfortable disconnect if healthcare is contributing to community or environmental harm, and should strive to minimize its environmental footprint. Hospitals are environments for healing, but contrary to this mission is the purchase of products and materials that are harmful to patients, staff and the health of their community. In order to move the supply chain, it has been important to strategically aggregate the demand in the healthcare sector to send consistent signals to the market about the need for greener products. We have created those levers to help accelerate change in healthcare sustainability specific to EPP.

The first lever was the Healthier Hospitals Initiative (HHI), which was an invitation to all hospitals across the country to address health and environmental impacts in the sector. The Initiative asked hospitals to commit to take on one more of the six proven strategies to improve their sustainability performance. After three years, over 900 hospitals signed on and shared data to measure market movement. The measurable successes included spending 18 percent of their food budget on local and sustainable food, over $6 million on green cleaners, and $12 million on furniture without targeted chemicals of concern. The success of this initiative led us to keep the challenges going.

As mentioned before, Healthier Hospitals is now a program of Practice Greenhealth, and, after fine-tuning the challenges, we have two working groups of members to leverage the aggregate buying power of participating health systems to accelerate the transformation of the healthcare supply chain toward more sustainable products, technologies and services.

We are also part of a new for-profit B Corp called Greenhealth Exchange (GX), which aims to accelerate the adoption of new and existing products used by healthcare providers with environmentally superior design and production features. GX will aggregate the purchasing volume of its founding healthcare organizations to make environmentally preferable products more affordable for healthcare buyers. GX will screen products and catalyze the development of healthier products for positive change. GX will be a market accelerator with a portfolio of only environmentally preferable products meeting the criteria set forward.

Quality process, quality outcomes

Quality improvement and quality outcomes are driven by the quality of the process that produces the product or service.[3] The two principal drivers are the degree to which:

- The process is visible and incorporates features that can prospectively identify and optimize the function of the process.
- Leadership can use a process to prevent a problem entirely versus mitigating a problem downstream.

Sustainability is founded on the principle that prevention is superior to downstream mitigation, treatment or rehabilitation. A visible process combined with a sustainability-oriented mindset, empowers leaders to anticipate, prevent and reduce negative outcomes through process improvement. That principle may be applied to clinical quality processes as well as to EPP processes. For example, designing a transparent clinical process enhances the likelihood that the clinical team will prevent an error from occurring because there is active involvement and participation by team members. It also enhances the degree to which the team can continuously improve the process because it is visible to all participants.[4]

These same principles may be applied to EPP processes. For instance, the healthcare leader may decide to stop purchasing single-use devices, opting instead for products that may be reprocessed. This decision focuses upstream to reduce waste from the supply chain process because the purchasing process is visible and may therefore be continually improved over time.

THE FRONT DOOR TO SUSTAINABILITY

A successful environmental sustainability program will not only consider how to dispose of waste responsibly, but will also look upstream to eliminate waste before it is generated. By choosing to bring materials into a facility that are created using less toxins, that do not emit volatile organic compounds, that use less energy and create less waste, hospitals can enhance healing environments and support healthy communities.[5]

Organizing concept: health, not just healthcare

Most purchasing decisions are incentivized by the prospect of short-term cost savings. The trouble with this approach is that short-term savings are not always aligned with the organization's long-term interests in community health. For instance, if Felipe, Vice President of Materials Management, is charged with

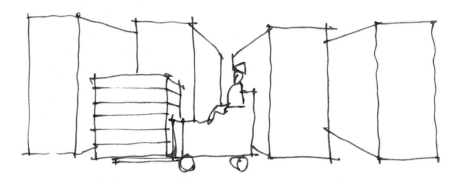

FIGURE 9.2 Warehouse of supplies

reducing departmental costs, he will likely base his purchasing decisions on cutting down immediate or first costs to reduce this year's budget. That approach contends with aligning Felipe's incentives around long-term savings and sustainability's triple bottom line.

Supply chain

The supply chain is a sequence of activities that takes place as raw materials are converted into products by manufacturers, purchased in bulk by distributors, and then purchased by providers in smaller quantities (Chopre and Meindl, 2004). EPP has many layers of complexity, beginning with understanding the impacts of a given product or service along its life cycle. A life-cycle analysis assesses the impact(s) products have at every step along the supply chain. EPP meets the supply chain at the outset of the supply chain process where decisions are made to incorporate environmentally preferable products.

Denise, Chief Financial Officer, and Felipe, Vice President of Materials Management, congregate in Denise's office to review Felipe's annual budget and cost-reduction targets.

FELIPE: I understand that I need to reduce our departmental costs in order to make budget. But I'm at a loss about how to make financially savvy decisions that don't undermine our sustainability goals. Would you please give me advice about how to balance short-term budget goals with long-term sustainability savings?

DENISE: [*Pauses while thinking how best to answer Felipe's question. Felipe gets a bit uncomfortable and fidgets in his chair due to the unusually long silence.*] I guess one answer is that you need to use metrics for both performance management and performance incentives, metrics that reflect long-term organizational goals. History is littered with organizations that managed quarter to quarter. These organizations made appropriate short-term decisions that positioned them for long-term loss. That situation is not only relevant to supply chain management purchasing decisions, but we're seeing a similar trend in other departments. For instance, during the recession, admissions fell and we laid off a number of employees, remember?

FELIPE: Yes, how could I forget?

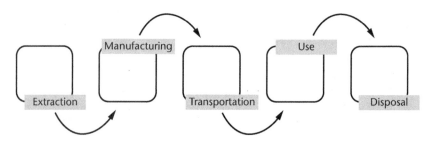

FIGURE 9.3 Supply chain process

DENISE: We may lag in recovering from the recession because we made poor short-term decisions. We sacrificed longer term and medium-term opportunities to gain more market share in the short term because we let go of experts who important during times when the economy is more robust. My guess is that it'll be hard to fill those positions with the comparable level of expertise, as evidenced before the economy tanked. Long-term consequences of our decisions often get lost in short-term thinking; that's the real risk of not thinking sustainably — we make short-term sacrifices that aren't in our community's long-term interests.

Knowing where to start

EPP can be separated into two key components, each having its own set of goals, tasks and measurement of progress.

The first EPP component is determining which products are purchased. Product purchasing with EPP in mind helps ensure that products we bring into healthcare settings do not pollute the environment during the product life cycle — from manufacturing, to the direct impact on patient health, through to the disposal process.

The second EPP component is exploring the EPP profiles of suppliers, ideally before contracting with them. The goal of this component is to work with suppliers to influence their sustainability-oriented procurement and manufacturing processes through the hospital's purchasing power. Think about this EPP component as a traditional background check: a leader does not hire an employee, consultant or vendor without researching the individual or business history. Why should suppliers be any different? The supplier background check should be complete before contract negotiation begins so that EPP can be worked into documentation. Please refer to the Appendix for a sample letter to a GPO requesting less toxic products. This letter was developed by the Minnesota Center for Environmental Advocacy's Polyvinyl Chloride (PVC) Task Force and provided by Minnesota Technical Assistance Program (MnTAP).

GLOBAL PERSPECTIVE

Many more hospitals in Europe are managed by government entities than in the United States, where hospitals are largely not-for-profit or for-profit.[6] Because of the ownership structure, various dimensions of healthcare management in Europe focus on longer term efficiency as compared with United States healthcare management models. European hospitals frequently receive global payments versus payment for specific services (WHO, 2011). In other words, they are economically incentivized to optimize prevention and reduce the utilization of healthcare restoration services. They are paid a fixed annual amount per patient, and if they can find cost-effective long-term purchasing strategies or facility development methods, hospitals save money on their global budget. The United Kingdom's National Health Service pays regions a fixed budget that is a capitated rate, similar to the Canadian health system (WHO, 2011). The power of this model is that both European and Canadian hospitals look upstream to impact prevention and reduce the utilization of treatments and facilities over the long term.[7] The European sustainability mindset extends from EPP processes to facility design, where the sustainability mindset is particularly visible.

Understanding the true cost of a purchase

Cost is one of the greatest perceived sustainability barriers. However, cost considerations often do not consider the total cost of ownership. The total cost of ownership includes all costs related to the extraction, manufacturing, transportation, use and disposal of materials throughout the supply chain.

EPP IN THE MARKETPLACE

To formulate an EPP strategy, it is important to first create benchmarks by evaluating and measuring current purchasing practices in all operational departments. This process provides a basis for identifying opportunities for which EPP policies and targeted goals can be developed. Opportunities exist to integrate EPP while working with existing contracts as well as while initiating new contracts. New contracts afford the opportunity to insert sustainability performance indicators. Such changes can be formalized by developing policies to support sustainability purchasing decisions in future contracts.

Practical application: Sweden's Industry Initiative for Sustainable Procurement

The imperative

- Achieve cost savings by capturing economies of scale.
- Create an information-sharing forum to lessen the learning curve within the industry.
- Pressure suppliers to improve the environmental and social impacts of health-care products.
- Promote the mission of healthcare to "first, do no harm."
- Reduce consumption today so that resources needed for continued operations are available in the long run.

The initiative

In 2006, a group of city and regional councils in Sweden conducted an impact analysis of quantity and dollars spent focused on regions where major healthcare products were produced and services procured. The councils used the results of the impact analysis – an examination of economic, ecological and social development issues within regions across the globe – to identify seven healthcare product categories that generally require closer monitoring up the supply chain due to their impacts:

1   Instruments and stainless medical products
2   Operating products, disposable
3   Gloves
4   Dressings
5   Textiles
6   Pharmaceuticals
7   IT.

Most of the products identified, as well as their component materials, are produced within Eastern and Southern Asia. Business, labor and environmental laws in these regions are less restrictive of practices that are prohibited elsewhere in the world. Healthcare organizations belonging to the city and regional councils that conducted the analysis support and promote sustainable procurement practices within these priority areas in two main ways:

•   Subscribing to a common code of conduct that addresses issues of environmental, ecological and social development up the supply chain; the code of conduct targets upstream impacts of priority procurement area manufacturers, parts suppliers and service providers.
•   Implementing a follow-up model for priority procurements, consisting of the following:
    •   Dialogue with suppliers: inviting existing and potential suppliers to discuss compliance with the code of conduct and review opportunities for improvement on an annual basis, if not more frequently.
    •   Observance of the code of conduct: requiring suppliers to show, on demand, that all components of the products and services they provide are produced in compliance with the code of conduct; for those that fail to meet the requirements, suppliers take corrective actions up to and including contract termination.
    •   Formal follow-up process: sending a self-assessment form to priority procurement area suppliers and assessing responses to determine where more detailed follow-up is needed; following up in further detail via requests for meetings with suppliers, remedial actions or live audits at the manufacturer; requiring an action plan for addressing environmental, social or quality issues, with occasional spot checks in production; inviting suppliers to seminars to review audit results and address any action required.

The metrics

Although it may be tricky at times to track the specific impacts of environmentally and/or socially preferable purchasing efforts, there are ways to monitor progress:

- Quantity of products targeted as priority procurements
- Quantity of and dollars spent on priority procurements
- Number of suppliers who have/have not met the criteria
- Specific product characteristics targeted to avoid (e.g., mercury-free, reduced packaging).

Lessons learned

- Several industry organizations have developed and made publicly available in-depth vendor questionnaires that specifically target certain chemicals, products and business practices. Using tools such as these vendor questionnaires helps standardize the sets of questions being asked of vendors from organizations across the industry, making it easier for them to understand exactly what types of information and products they are looking for.
- Since healthcare organizations share many of the same suppliers, there is great potential to standardize the industry's approach to procurement and purchasing. Working together with other healthcare organizations to implement vendor evaluations, supply chain investigations or industry-wide discussions makes accomplishing the task more impactful than when each institution works individually.
- Suppliers are often national companies with subcontractors abroad. When conducting a supplier or product analysis, follow-up must extend up the entire chain of production to ensure that comprehensive supply chain impacts are identified and considered.
- Certain product groups are more vulnerable than others to environmentally or socially harmful practices during the production process. Start small by developing an environmentally preferable purchasing program that targets these product groups, rather than trying to build a comprehensive program from scratch.

About the organizations

Since 2006, Stockholm County Council, Skåne Regional Council and Region Västra Götaland have been working together to promote social responsibility and sustainability in procurement. Together, the group developed a common code of conduct and follow-up model for suppliers. The county councils and regions in Sweden are primarily the principal authorities for healthcare, and also for public transport and sustainable growth in their respective regions.

Sources

Stockholm County Council et al. (2010). *Social Responsibility in Procurement: Code of Conduct for Suppliers*. Sweden: Stockholm County Council, Skåne Regional Council and Region Västra Götaland.

www2.vgregion.se/upload/Regionservice/Ink%C3%B6p/H%C3%A5llbar%20
upphandling/Social%20responsibility%20in%20procurement%20-%20Code
%20of%20conduct%20for%20suppliers.pdf.

Leadership imperative

Like every other dimension of sustainability, EPP points to leadership and leadership's role in promoting sustainability as a core component of value for the organization. Creating a successful EPP program is contingent upon leadership. Leadership
is crucial to having EPP within an organization, and is equally imperative when
facilitating EPP relationships with industry partners. It is the leader's role to make it
clear to individuals in decision-making positions that sustainability and prevention
are core values. It is the leader's role to empower personnel at all levels of the organization to make purchasing decisions based on life-cycle cost and triple-bottom-line
sustainability criteria. This chapter highlights some of the challenges and solutions of
EPP, both of which are morphing as the EPP industry develops.

Fred and Felipe are conversing in the hallway before a meeting begins.

FELIPE: Hi, Fred. Good news – we've instituted a hospital-wide EPP policy and it's
really starting to show results. We're saving energy because we bought the
environmentally preferable lights and now our engineering folks are delighted
– fewer light bulbs to change over the long run. The operating room nurses
are happy too because we've been using EPP to source the intravenous fluid
management system. Now the nurses have minimized exposure risk because
they no longer need to handle bodily fluids – the machine takes care of it all
while not throwing away a bunch of canisters for every OR case, the way we
used to before we purchased the fluid management system.[8]
FRED: I'm starting to feel good that we did the EPP thing. [*Fred's phone rings. It's
Mrs. Crumpmuffet, Fred's assistant. Mrs. Crumpmuffet tells Fred that the Mayor is on
the phone. Fred speaks into his cell phone.*]
FRED: [*to the Mayor*] This is Fred speaking.
MAYOR: Good morning, Fred. This is the Mayor here.
FRED: Good morning, ma'am. What can I do for you today?
MAYOR: Well Fred, I understand that Memorial Hospital purchased new lighting
technology that can be retrofit, which ultimately saves energy for the com-
munity. I would appreciate if you would come by so that we can discuss those
projections. I am delighted to learn more.
FRED: [*chuckling to himself*] Thank goodness I saw the light before the Mayor
showed up.

Leadership dialogue

1   Quality and environmentally preferable purchasing share common underlying
principles, and leaders often find resistance to both topics. What are the causes
of resistance?

2 Why do or don't leaders have to make a choice between healthcare (retroactive; healing the sick) and health (proactive; preventing harm)?

3 Why is the "true cost" of purchasing situation-specific?

4 When the leader brings an environmentally preferable purchasing proposal to the Board of Trustees, it may be subject to criticism because of the short-term financial impact upon the institution. How can a leader help the Board understand the true cost to the hospital of short-term thinking versus the long-term perspective associated with sustainability?

Syllabus integration

Overview

We will examine best practices for preferred purchasing that document efficacy, infection control, environmental and human life-cycle impacts from manufacture to waste, as well as effective capital and maintenance costs.

Objectives

1 Evolve supply chain management from a traditional focus on purchasing and logistics to a broader, more integrated emphasis on contributing to value creation:

a Identify and illustrate opportunities to enhance supply chain performance by creating business value in the procurement supply chain.

2 Study how sustainable supply chain management affects business and relationships with manufacturers, suppliers, clinicians, managers and practices:

a Detail how companies in other industries derive business value from sustainable sourcing.

3 Understand how cleaning practices support prevention and control goals.

4 Link occupational health to healthcare sustainability.

Topics

1 Drivers of environmental health and safety successes

a National Institute for Occupational Safety & Health (NIOSH)

2 Key questions of the environmental health and safety supply chain

3 Environmentally preferable purchasing

a Group purchasing

b Business process improvements

c Environmental and social benefits

Homework

Healthier Hospitals (2017). Smarter purchasing. Available at http://healthier hospitals.org/hhi-challenges/smarter-purchasing.

Practice Greenhealth (2017). *Environmentally Preferable Purchasing.* Available at https://practicegreenhealth.org/topics/epp.

Additional resources

Kaiser Permanante (2011). *Green Cleaners.* Available at http://practicegreenhealth. org/sites/default/files/upload-files/environmentalservicesgreencleanerssuccess story11111.pdf (accessed October 22, 2012).

Practice Greenhealth, Greenin the OR Case Study (2011). *North Suburban Medical Centre: Fluid Management in the OR.* Available at http://practicegreenhealth.org/ sites/default/files/upload-files/casestudy_nsuburban_r5_web_0.pdf (accessed October 22, 2012).

Practice Greenhealth, Greenin the OR Case Study (2011). *The University of Maryland Medical Centre: Reusable Textiles in the OR.* Available at http://practice greenhealth.org/sites/default/files/upload-files/casestudy_uofmd_r6_web_0.pdf (accessed October 22, 2012).

US Department of Labor – Occupational Safety and Health Administration (OSHA). *Healthcare Hazards.* Available at www.osha.gov/SLTC/healthcare facilities/otherhazards.html.

Notes

1 NPHPPHC (2011). *National Prevention Strategy: America's Plan for Better Health and Wellness.* Rockville: National Prevention, Health Promotion and Public Health Council. Available at www.surgeongeneral.gov/priorities/prevention/strategy/report.pdf.
2 The Global Importance of More Sustainable Products in The Global Health Care Industry, a study commissioned by Johnson & Johnson.
3 Hughes, R. (n.d.). Tools and strategies for quality improvement and patient safety. Agency for Healthcare Research and Quality. Available at www.ahrq.gov/qual/nurse shdbk/docs/HughesR_QMBMP.pdf.
4 Casey, J. (2007). *A Lean Enterprise Approach to Process Improvement in a Health Care Organization.* Massachusetts Institute of Technology. Available at https://dspace.mit.edu/ handle/1721.1/42345.
5 O'Brien-Pallas, L., Griffin, P. and Shamian, J. (2006) The impact of nurse turnover on patient, nurse, and system outcomes: a pilot study and focus for a multicenter international study. *Policy Politics Nursing Practice* 7(3): 169–79.
6 Health Forum LLC (n.d.). Fast facts on US hospitals. American Hospital Association. Available at www.aha.org/research/rc/stat-studies/fast-facts.shtml.
7 Klein, E. (2012). Why an MRI costs $1,080 in America and $280 in France. *Washington Post*, March 3. Available at www.washingtonpost.com/blogs/ezra-klein/post/why-an-mri-costs-1080-in-america-and-280-in-france/2011/08/25/gIQAVHztoR_blog.html.
8 Practice Greenhealth (2011). Greening the OR: fluid management systems in the OR. Available at http://practicegreenhealth.org/webinars/greening-or-fluid-management-systems-or.

10

BUILDING BRAND, LITERALLY

Growth/brand + green building

FIGURE 10.1 Building green buildings

Executive summary

Leaders increasingly recognize the financial, environmental and social benefits of pursuing sustainable building retrofits and new construction. Among the drivers of sustainable decision-making is the potential for long-term reduction in life-cycle costs. Healthcare institutions are embracing commitments to environmentally friendly building design by pursuing Leadership in Energy and Environmental

Design (LEED) standards and LEED for Healthcare, which incorporates the unique rules, regulations and operating conditions of sustainable healthcare environments. The impact of using LEED standards in healthcare facilities has important implications for the brand and brand elements that are communicated through facilities. Sustainability is a desirable brand element for employees and the various publics served by healthcare organizations. This chapter describes the sustainable facility design guidelines that align leadership priorities with environmental health practices in the planning, design and construction of facilities, as well as their impact upon the institution's brand.

Storyline

Memorial Hospital conducted a market analysis to identify growth opportunities. It determined that building a cancer center would fulfill unmet community needs. Fred discusses his vision for the cancer center with Melvin Monroe, Vice President of Design and Construction, and Fiona O'Malley, Director of Marketing and Communication. Their conversation quickly turns to the importance of making sustainability visible by intentionally integrating it into the building design and organizational brand.

MELVIN MONROE: Witnessing my father's recent cancer bout caused me to reflect on facility design through a whole new lens. I observed that the construction materials used in the cancer center include materials that have been judged to be carcinogenic. Can you believe how ironic that is?

FIONA O'MALLEY: I'm so sorry to hear about your father.

MELVIN MONROE: Thank you, Fiona. There's nothing I can do about my father's situation now, but there is something we can do to improve the experience for other people's families. I think we need to be explicit about our commitment to sustainable design with this new cancer center.

FRED: I'm not sure what it would mean for us to commit to sustainable facility design, but however we incorporate it, sustainability needs to be visible if we want it to resonate with employees and patients. Sustainable design isn't helpful from a brand perspective unless we translate it into tangible features. Good intentions are not good enough. We have to have concrete sustainability characteristics that we incorporate systematically into the facility design.

MELVIN MONROE: You want to design the facility out of concrete?

FIONA O'MALLEY: Oh, Melvin, that joke was awful.

FRED: What I was trying to say is that we should make our sustainability features obvious to even the casual observer. And then we have to point out the sustainable features to people so they know what they're looking at. The building itself can be educational wherein the medium is the message.

MELVIN MONROE: This will be our first building designed to sustainable standards. I'm looking forward to this project.

FRED: How will people know if we invest in sustainability? For instance, how will the community know if we invest in sustainable transportation or waste management systems? Patients are unlikely to be poking around the dumpster exclaiming, "Oh, now this is a sustainable hospital!"

MELVIN MONROE: There are things we can do to indicate our sustainable commitments to the community, even for sustainable features that are less visible. For example, we could incorporate signage in the hospital facility to highlight energy conservation or on the side of the shuttle bus, to highlight our reduced carbon footprint.

FRED: Some people might say that they don't need to care about the substance to care about the perception, but I find myself caring more as I think about my responsibility as CEO in an increasingly sustainable industry. I care a lot about how sustainable the building looks. I'd love to have a garden growing on the roof, with landscape design that recycles water. Solar panels would make our commitment visible. That's how we build a sustainable brand: by investing in and showcasing our sustainable features. [*Fred pauses, then grins.*] Feeling green today, Melvin?

MELVIN MONROE: Green with envy, sir.

EFFECTS OF THE BUILT ENVIRONMENT

The US EPA created the chart in Table 10.1 to articulate the effects of the built environment on environmental and human health. This reality – that buildings impact public health – is not a typical leadership consideration, though it should be.

TABLE 10.1 The impact of buildings upon health[1]

Aspect of built environment	Consumption	Environmental effects	Ultimate effects
Siting	Energy	Waste	Harm to human health
Design	Water	Air pollution	Environment degradation
Construction	Materials	Water pollution	Loss of resources
Operation	Natural resources	Indoor pollution	
Maintenance		Heat islands	
Renovation		Stormwater runoff	
Deconstruction		Noise	

Brand new impact

Every organization has a de facto brand. Healthcare organizations are no exception. Hospitals tangibly articulate organizational brand when they invest in facilities and cutting-edge technologies because such assets are visible to the community.

WHAT IS BRAND?

Many of today's healthcare facilities are outdated, having been built during the Hill–Burton (1946) era.[2] Consider the facilities of forward-thinking, successful businesses. Think Google. Think Amazon. Think Apple – "Think different." These companies are explicit in their sustainable design, ranging from facility design to product design. They purposely communicate that the company embraces sustainable values. For instance, Apple recycles used products; Whole Foods makes an effort to source local foods. Both companies streamline their services to provide a convenient, minimally wasteful customer experience. Companies that intentionally brand themselves attract clients, customers, suppliers and creative talent with shared values. Brand is one of several advantages accruing to hospitals that choose sustainable building design. Having a visibly sustainable building is, at a very practical level, one durable strategy for building sustainability into the hospital's brand.

WHAT IS A SUSTAINABLE FACILITY?

Sustainable design is the process of incorporating products and processes that meet the triple bottom line. Sustainable facilities are environmentally, socially and fiscally responsible in the long term. They are responsible stewards of community resources, including land, water and energy, while reducing waste and chemical use.

One of the most prevalent sustainable building systems is the United States Green Building Council's (USGBC) Leadership in Energy and Environmental Design (LEED). LEED for Healthcare was designed to incorporate healthcare's unique challenges, such as 24/7 operations and air-quality management. The LEED for Healthcare rating system evolved from the Green Guide for Healthcare, a pilot rating developed by industry leaders.

OTHER THIRD-PARTY GREEN BUILDING RATING SYSTEMS

Include:[3]
- Australia Greenhouse Building Rating (AGBR)
- BCA Green Mark (Singapore)
- Building Environment Assessment Method – Hong Kong (HK – BEAM)
- Building Research Environment Assessment Method Consultancy (BREEAM) (UK)
- Comprehensive Assessment System for Building Environment Efficiency (CASBEE) (Japan)

- Ecology, Energy Saving, Waste Reduction and Health (EEWH) (Taiwan)
- Energy Star (US EPA)
- European Environment Agency
- Green Building Council (Korea)
- Green Globe Rating System (United States)
- Green Star (Australia)
- India Green Building Council
- Leadership in Energy & Environmental Design – Canada (LEED – Canada)
- Leadership in Energy & Environmental Design (LEED – United States)
- Philippine Green Building Council

Build it, and they will come

PHYSICIANS AS CUSTOMERS

Physicians are key hospital customers because most patients go to the hospital as directed by their physicians. Thus while a hospital's direct-to-consumer (patient) brand is important, leaders cannot ignore the fact that only physicians admit patients, other than emergency room visits. Brand does, however, play a key role in elective procedures and for employee recruitment and retention.

How does a leader attract top talent in a tight market? The "sweet spot" in brand differentiation is linking product and service differentiation with talent recruitment. Patients, physicians and employees seek care and work at organizations that have visible, explicit commitment to shared values. A sustainable facility environment promotes the health of patients and employees, which is an attractive component of the organization's brand promise. Sustainable building design contributes to an organizational brand that attracts people who share sustainability values. While sustainable buildings are necessary, it is even more important to operationalize sustainable work processes and a culture of sustainability within the hospital walls. In other words, it is necessary but insufficient to incorporate sustainable facility elements that do not have an impact upon work processes or culture.

Does LEEDing cost more?

Questions about cost implications of sustainable design typically arise once leaders recognize that sustainable facilities are a desirable option. Leaders typically ask, "Does building according to LEED standards cost more than building less sustainable solutions?" There are gaps in most leaders' understanding of the somewhat complex answers to this question. There is considerable cost variation inherent in the planning and construction of sustainable buildings for a number of reasons. There are different levels of LEED standards that range from the most sustainable (platinum) to the least sustainable (certified) level of LEED. Sustainable features may, in aggregate, cost less or more to build to LEED standards, depending on which line items within the LEED rating system are incorporated. In general, the initial investment in high-level LEED standards is higher than the initial investment in low-level LEED standards.[4] Upfront construction costs for healthcare facilities average 1.24 percent higher to build according to LEED standards than to not do so (Vittori et al., 2013).[5] The question then emerges: is building sustainably "worth it" when the incremental building cost is spread over the useful life of the building, assuming the average useful life is 30 years?

Life-cycle cost analysis consistently demonstrates a healthy financial return on incremental capital investment owing to energy and other operating cost savings.[6] In other words, the incremental difference in building cost for sustainable buildings is relatively small due to the long-term return on investment. But the question is not solely that of incremental building cost. Rather, what is the overall value equation that is determined by comparing incremental costs with the incremental benefits that result from the sustainable building in terms of patient, employee, physician and public perception? It is a question of value, not only expense, that should be considered.

Green building design has the opportunity to impact the health of occupants, including healthcare staff. (Huang, 2011). In addition to quality and health outcomes, patient satisfaction is of key importance in healthcare. The ability to build a space that can appeal to patients' satisfaction is an added benefit (Allen et al., 2015).

Buildings deliver on the brand promise

The life of Steve Jobs taught the world that exceptional design incorporates both functionality and the aesthetic features of products, services and companies.[7] Applied to healthcare, an architecturally striking cancer center communicates cutting-edge care. In other words, the facility design communicates an important element of the brand promise. Jobs made his mark by knitting together the left brain and right brain.[8] He captured the left brain's reliability, safety, consistence, quality and durability, and married that with the right brain's ability to capture the style and creative design for which Apple products are known.[9] Sustainability incorporates the evidence-based methodologies that yield predictable, proven results. Sustainability also requires the creative, right-brain innovation which is reflected in trying out new methods to achieve a triple bottom line. High-performing leaders in healthcare have the ability to apply both left- and right-brain thinking ambidextrously. Leaders who embrace sustainability embrace that approach to accomplish their organization's mission: improving the long-term health of their patients, their community and the planet.

Fred and Fiona hold a visioning session for a marketing commercial being produced for the new building.

FRED: We can't just claim that Memorial Hospital is environmentally sustainable. If we promise A but deliver B, we're in trouble. It is essential that we practice sustainability or it will be readily apparent that our claim is insincere.

FIONA O'MALLEY: [*Director of Marketing and Communication*] I understand what you're saying, Fred, and as the person responsible for brand management, I fully agree with your point. Based on the dialogue I've been hearing, perhaps we should incorporate sustainability as an explicit part of our brand – from our workflow processes to our physical infrastructure. What do you think?

FRED: Are there healthcare organizations of our size that explicitly tie sustainability to their brand?

FIONA O'MALLEY: There are healthcare organizations that practice sustainability, but no organizations of our size explicitly link sustainability to their brands. [*A pause of silence overtakes the room as Fred and Fiona consider brand options.*]

FIONA O'MALLEY: I've been thinking: we can change what something costs and we can change how something looks. That's why leaders who are experts on cost-effectiveness will plug cost-effectiveness everywhere they touch. Similarly, sustainability-focused colleagues plug sustainability everywhere they touch. But the most influential leaders are those who protect the values nobody can change, the values that are fundamental to the hospital's culture. Once we build specific values into the culture, those values become the organization's brand. At that point, customers see our brand values as differentiators of Memorial Hospital. So, if you really believe that sustainability is a brand differentiator, we need the culture to demonstrate it. [*Fred stares out of the window, envisioning a brand defined by sustainability values.*]

FRED: Healthcare organizations are increasingly associated with improving health, not just healing the sick. Inherently, then, sustainability is what our brand is moving toward. Our new brand is about looking upstream to prevent illness, helping patients avoid the hospital in the first place. At its core, the principle of predict-and-prevent is the future paradigm of health.

FIONA O'MALLEY: Our new brand is inherently about being more sustainable and sustaining oneself, but I don't know that a lot of people would connect the dots unless we make it obvious. Sustainable facilities would illustrate our commitment to sustainability.

FRED: I agree with you. There may be only two people in this hospital who reach that conclusion. [*Sustainability Director Amalie DuBois enters Fred's office, where she is greeted by Fred and Fiona.*]

AMALIE DUBOIS: Good afternoon, Fred. Good afternoon Fiona.

FRED: Hello, Amalie. Fiona and I were just talking about how excellent it would be for our facilities to convey the sustainability brand we aim to achieve. If that was easy and obvious, everybody would be doing it. I think we probably have an answer for how we should do it, I just haven't thought of it yet.

AMALIE DUBOIS: Let me guess, you'd like for me to think about it.

FRED: That's right. You're good. You must be a mind-reader, Amalie. What am I thinking now?

AMALIE DUBOIS: You're thinking you want to give me a raise [*laughter*].

GREEN LEASES

Green leasing is slow to be adopted in healthcare as a general practice. Green leases are precisely what the phrase implies: leasing space that incorporates sustainable practices into legal documentation.[10] The lease is explicit that the real estate management company responsible for managing the property uses vendors and processes committed to sustainability, such as waste management services that include recycling. Demand for green leasing is likely to increase as businesses focus on sustainability and integrate it into operational practices. For healthcare, this trend means that green leases will grow increasingly popular in outpatient clinics, physician offices, medical office buildings, corporate office space, data management centers and other leased healthcare spaces.

Healthcare facilities carry out meaningful roles besides brand building and their inherent functional purposes. They serve to educate and are agents of culture change within the community. One valuable aspect of LEED certification is signage that explicitly calls attention to sustainable features. Examples of educational signage include signs indicating that the cafeteria recycles, the toilet uses less water than traditional options, the carpet contains recycled content, the lights turn off automatically and the healing garden water is gathered from rain barrels. Patients and employees are reminded that it is possible to incorporate these elements into a healthy home environment. Sustainable facilities clearly inform visitors, employees and patients about how to live more sustainable, healthful lifestyles.[11]

Core values

FRED: Our employees ought to be our brand ambassadors, ambassadors of sustainability.

FIONA O'MALLEY: I think you're on to something. [*Fiona notates furiously on her iPad.*]

FRED: We need to first educate employees about why sustainability is important to the hospital and how we embrace sustainability. The problem is that we're talking about things Memorial Hospital could do, probably not things we currently do. I can think of a million messages [...] [*Fred's voice drifts off as he mumbles brand messages.*]

FIONA O'MALLEY: Of all of the competing messages, let's pick one core message to focus on.

FRED: Okay, how about prevention? Come to Memorial Hospital and we'll help you prevent disease.

FIONA O'MALLEY: That's a possibility. Something like that would allow us to fold sustainability into the brand message, though perhaps not explicitly.

FRED: Gee, Fiona. I really hadn't thought about sustainability in this light until we started discussing it. I'm realizing that the environment we create speaks volumes about whether sustainability is perceived to be part of our core values.

FIONA O'MALLEY: That's right. We need to be very careful about the way we build brand into our permanent facilities. Our buildings articulate organizational values to the community long after you and I retire.

FIGURE 10.2 Watching growth

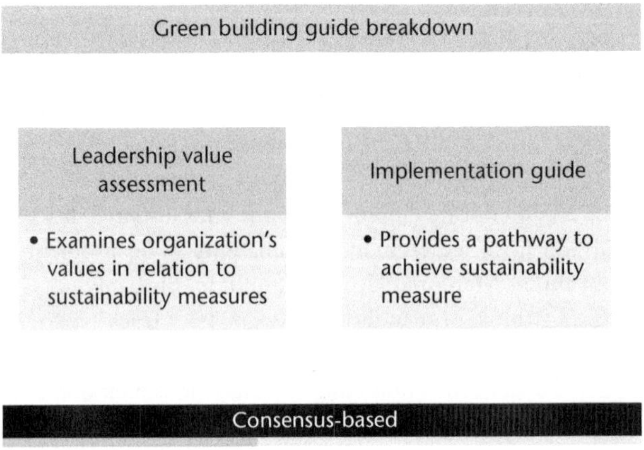

Green building guide breakdown	
Leadership value assessment	**Implementation guide**
• Examines organization's values in relation to sustainability measures	• Provides a pathway to achieve sustainability measure
Consensus-based	

FIGURE 10.3 Leadership value assessment

Internal guidelines

Let us assume that a leader decides to embrace the physical environment as a way to integrate organizational values around sustainability. LEED offers an array of guidelines, primarily for design, architecture, construction, engineering and related industries. Internal building guidelines may also be used by leaders to translate sustainability values into actual building projects. Kaiser Permanente is one example of an organization that relies on internal facility guidelines (Kaiser Permanente, 2011). Inova Health System has also developed internal guidelines for sustainable facility standards.

Brand building requires the promotion and communication of core themes or features. The features must be designed so that they support delivery of the brand promise. In other words, it's not enough to know that you want sustainability features integrated throughout the building. The design team needs to understand what "sustainability features" mean and how to incorporate them. That includes a system of quantified goals and metrics to evaluate progress, as well as tools to measure the impact of sustainability practices and provide ongoing feedback about how to continuously improve.[12]

FRED: We ought to have sustainable building principles to articulate the character of our facilities. A logical place to start is the major renovation project coming next year. We need to focus on both substance and appearance for our sustainability principles, everything from vendor partnerships to signage.

MELVIN MONROE: I can't help you much on the vendor partnerships piece, but we can certainly develop sustainability principles for our construction projects. We just had a new colleague named Dwight join the team who's a recent college graduate and already passed the LEED exam. Apparently, learning and practicing sustainably is the way people are trained these days.

FRED: At this point in my career, I think I'd be incompetent at practicing sustainability. I'd have to get Dwight to mentor me. If he can educate me about sustainability, anybody can be educated, right?

MELVIN MONROE: You said it, not me, sir.

> Some 99.1 percent of respondents answered that environmental sustainability ("Going Green") is important to them, but only 83 percent are satisfied with Inova's commitment to Going Green (Practice Greenhealth, 2012).

It's who we are, not what we do

Sustainability is often a bolted-on external feature that goes something like this: "We got the budget down. We designed the building. We got the program ready. Did anyone apply for the LEED-certified thing? We get something if we do that." In healthcare, sustainability is still an "externality" that few organizations have fully integrated into how they think or, more importantly, who they are. Maturity will be evident in healthcare when sustainability comes from the inside out, and not the reverse.

FRED: Every time we have one of these conversations, I go back with a long list of things I need to do to help us think and behave more sustainably. Talking it through helps shed light on how relatively early we are on this sustainability journey. We are still in our infancy, still wearing diapers, as far as sophisticated sustainability organizations are concerned. There's something to be said for sustainability education.

FIONA O'MALLEY: Just imagine, Fred. Once we integrate sustainability into our core caretaking processes, we can begin to integrate sustainability into the community in more meaningful ways. Let's say we're going to send a new mother home with her baby. Today, we teach the mother about breast feeding; we teach her about how to put the baby in the crib safely. Part of our new parent training could also entail helping parents create healthy, sustainable environments for their newborns. That includes how to maintain the home environment in a sustainable way. We need to build sustainability principles around the parent's obligation to optimize health for his or her child. Everything from the absence of lead paint in the home to the way diapers are disposed.

FRED: Enough diaper talk. Who's ready for lunch? In all seriousness, I do get what you're saying. We could organize our services around the concept of health, which is a friendly concept to parents, and a way to create sustainability ambassadors of our patients and families. I think it's a great way to get the nursing staff more involved, too.

OUR ENVIRONMENT AND OUR HEALTH: A CLIMATE CONNECTION

According to a recent study produced by *The Lancet*, one of the world's leading medical journals, "tackling climate change could be the greatest global health opportunity of the 21st century." While climate change is not a new issue, there is an emerging opportunity to address climate change through a new lens. Historically, the conversation has been deeply rooted in politics. The new vantage point is built on the foundation that climate change is a healthcare issue. The new spokespeople of climate change come with a background in health. The voice of healthcare leaders will help amplify climate change as a health issue. Addressing the related health impacts of climate change will be a paradigm shift that depoliticizes the issue to address it in a way that is relevant to all.

Climate change is a socially charged issue. Until recently, climate change has largely not been addressed by the healthcare industry. There are existing resources and studies about the impacts of climate on health; however, there is a gap in making the information accessible and actionable for the healthcare industry.

SUSTAINABILITY IN ACTION

Gail Lee is Sustainability Director at the University of California, San Francisco. She brings ten years of environmental health and 12 years of healthcare experience to this position to lead UCSF campus and UCSF Health on its sustainability journey. Her overarching challenge is to achieve the University of California's policy goal of Carbon Neutrality by 2025 and Zero Waste by 2020 while aligning with UCSF's mission of Advancing Healthcare Worldwide™.

UCSF brings together the world's leading experts in virtually every area of health. We are home to five Nobel laureates who have advanced our understanding of cancer, neurodegenerative diseases, HIV/AIDS, aging and stem cell research. UCSF Medical Center, UCSF Benioff Children's Hospitals, all four of its professional schools – dentistry, medicine, nursing and pharmacy – and many UCSF graduate programs consistently rank among the best in the country, according to the latest surveys by U.S. News & World Report.

What do you think the market is going to do?

Sustainability is definitely a focus in higher education, but it has not been a priority at most hospitals across the country. Many are beginning to recognize the value of energy efficiency, water conservation and waste reduction as avenues to cost savings and good community health. Currently, the leaders in this area are Gundersen Health System, the Cleveland Clinic and Kaiser Permanente. Practice GreenHealth, the nation's leading membership and networking organization for the healthcare community, has made a commitment to share sustainable, environmentally preferable practices with its 1,600+ members and to continue to increase its impact nationally.

What growth have you seen?

I have observed the membership increase over the past decade of Practice Greenhealth, Hospitals Without Harm and the Healthy Hospital Initiative. It's heartening to see the increased participation in the Practice Greenhealth Award application that collects data for benchmarking and provides direction for hospitals to become more sustainable, reduce their environmental impact and save money. More can be done to increase a focus on energy efficiency in this industry that contributes to 8 percent of GDP and generates enough greenhouse gas emissions to land itself among the top emission-producing countries in the world.

Why did you take on this project? What inspired you? (Was it cost savings, co-benefit with another project, employee engagement benefit, etc.?)

We recognized that sustainable efforts protect environmental and human health. An article in *The Lancet* states: "Climate change underpins all the social and environmental determinants of health," and "Responding to climate change could be the greatest global health opportunity of the 21st century." We engage our community by tying sustainability to UCSF's mission of *Advancing Health Worldwide*. Providing employee education, engagement and recognition drives more support for our programs. Our LivingGreen office/laboratory/clinic certifications engage staff. Our team helps measure energy and waste savings by using timers, power strips, waste-sorting education, and by providing outreach events such as our poster series on Climate Changes Health as well as employee recognition at our annual Sustainability Awards Ceremony. We also provide employee discounts for residential solar panels, electric vehicles and green retailer coupons.

Who are the participants in this project?

Our Academic Senate Sustainability Committee, the UCSF Advisory Committee on Sustainability and the Sustainability Steering Committee are all volunteer middle to upper management members or faculty. We also have a mostly volunteer staff who make up nine workgroups that implement sustainability projects in the areas of climate/energy, water, waste, procurement, green building, green operations, sustainable food, toxics reduction and culture shift. Other grassroot efforts are run by volunteer staff, students and faculty who embrace their own projects. In the past two years, we have hired two Bay Area Climate Corp fellows who have provided ten months of full-time support to help with sustainability project implementation.

What cost savings has UCSF experienced as a result of the organization's sustainability efforts?

We reduced our waste management costs by educating and encouraging waste segregation. We reduced landfill waste and increased recycling and composting. A recent campus waste-sorting grant increased our diversion rate to 80.1 percent, generating enough cost savings to hire two full-time employees to keep the program going. The biggest savings occurred at UCSF Health in the reprocessing of invasive and non-invasive medical devices, diverting 62,000 pounds and saving US$1.7 million in 2016 through a 50 percent discount on the repurchase of those devices. Over the past six years, energy-efficient projects through a partnership with our local utility have helped us achieve an annual cost avoidance of US$1.5 million for UCSF Health and US$2.7 million for UCSF's campus.

What are the key outcomes you have witnessed as a result of the sustainability program and what are the positive impacts upon the community?

The cost savings have been paramount to leadership, which gives us some cache, institutional value and support to continue our work. Faculty and students participated in incorporating climate change and health into the curriculum, which has reinforced the message more deeply in the academic arena. Our staff are proud of their accomplishments and the recognition we and they receive. Tying sustainability to the university's mission and providing recommendations for action appears to be taking hold. Each year, more and more engagement is seen across the organization.

What are the resulting health benefits?

Health benefits can be measured through a drop in the rate of respiratory disease such as asthma, vector-borne and water-borne infectious diseases, extreme heat-related deaths and pre-term births. We've measured reduced air pollution and CO_2 emissions from fossil fuels used for energy production and transportation coupled with avoidance of landfill methane production through composting. These combined efforts lead to improved air quality and human health benefits, ultimately resulting in healthier communities that directly align with UCSF's mission of Advancing Health Worldwide™.

If you were to embrace sustainability all over again, what would you do differently? What are your lessons learned?

We would earn leadership support by first focusing on projects with the biggest cost savings. At the same time, we would tackle grassroots projects through staff interest in recycling and waste reduction, which is more visible and engaging for the frontline employees. We've learned that a two-pronged approach, focusing from the top down and the bottom up, is most effective. Seeking increasingly more engagement from the UCSF community is also a priority to ensure we get cooperation from the public for the sustainability programs we want to implement in the future.

How did you measure the success of your sustainability program? Is it related to cost, quality, engagement, etc.?

It's measured in multiple ways through our software tool which tracks emissions from electricity and natural gas, all our waste streams and diversion rates, water, shuttle and vehicle fleet emissions, refrigerant releases, medical gas usage, and commute and business travel emissions. We also track the number of people engaged at sponsored events and outreach campaigns and a survey that measures the UCSF community's perceived impact of our efforts upon their behavior.

How did you convince others to get on board with the sustainability program?

We engaged the University Chancellor to recognize individuals who made significant contributions to sustainability at an Annual Sustainability Awards

Ceremony. Our monthly newsletter shares our successes and encourages our community to provide support. In addition, we participate in tabling at other events; for example, we created a poster series campaign to link climate change and health. We continually seek other avenues of engagement.

What tools and/or resources have you found helpful in your journey?

We provide tools such as the office/lab/clinic certifications, energy meters, light meters, timers, posters, employee discounts, drawings prizes, tabling events such as our LivingGreen Fair/Bike to Work Day. Working in partnership with the City and County of San Francisco, our utilities and our waste haulers has contributed to our success.

What is your advice for the next person/organization that would like to embrace sustainability?

Be engaging and appeal to healthcare providers' altruism. Healthcare leaders are in this industry for a reason – they care about people. Be positive and have measureable results. Report all successes back to the community and share best practices. Recognize the efforts of others and tie sustainability work back to the mission of the organization.

Practical application: green buildings across the globe

The imperative

- Reduce employee, patient and visitor exposure to chemicals of concern.
- Improve patient outcomes.
- Improve employee performance and satisfaction.
- Reduce workplace injuries and employee health costs.
- Achieve long-term savings through energy and water efficiency, reduced waste, and lower operations and maintenance costs.
- Create environments that promote healing.
- Minimize long-term risk related to energy supply and prices.
- Promote public health by reducing environmental impact.

The initiative

The world's first two LEED Platinum hospitals – Dell Children's Medical Center in Austin, Texas, USA and Kohinoor Hospital in Mumbai, India – showcase the universal benefits of green building, despite being located across the globe from each other (Dell Children's Medical Center, 2008; Express Healthcare, 2011). Each building has capitalized on the unique features of its physical surroundings to create a healing environment that not only promotes environmental responsibility, but also yields significant financial savings, increased employee productivity and retention, and faster healing times, among other benefits.

LEED is an international certification program for green buildings, which rates buildings in five key areas. Dell Children's Medical Center and Kohinoor Hospital were designed to be top performers in each of these areas. Both hospitals undertook some of the same green building activities in each category – practices that tend to be universal throughout the global green building sector – while each organization also had unique ideas suited to their specific sites.

- Both organizations used recycled and original site materials in the construction of their new buildings. Dell was located on the site of a former airport and used the original runway materials in its parking lots and garages; Kohinoor used re-rolled steel and recycled frame wood in its exterior façade, flooring and walls.
- Both organizations focused on increasing the amount of natural light in indoor areas, which they achieved through different mechanisms – Dell through interior gardens and courtyard spaces; Kohinoor with large outward-facing windows and open skylights.
- Both organizations committed to energy efficiency, approaching the topic in different ways. Dell used strategically placed air intakes, built a natural gas-fired power plant (75 percent more efficient than coal), and installed an underfloor air-distribution system requiring less fan power than traditional ceiling systems. Kohinoor installed solar hot water heaters, procured green power from windmills, and focused on improving heating, ventilation and air-conditioning performance by installing new, highly efficient systems.

The metrics

The design and construction team can help track building performance throughout the construction process, both on existing buildings and on new construction projects. Commonly tracked metrics include:

- Quantity of construction and demolition waste reused or recycled.
- Amount of sustainable and/or recycled materials included in building.
- Total dollars spent on locally produced materials.
- Total cost and environmental savings related to improved energy and water efficiency (gallons of water reduced or pounds of CO_2 eliminated).
- Quantity of chemicals avoided in building materials.

Lessons learned

- The most successful green building projects tailor common green building practices to the physical (site, climate) and social (regulatory environment, business culture) conditions of the region.

- Investing in green buildings takes only a small increase in initial cost over conventional buildings while the benefits are long term and multifold. The appropriate expertise and planning must occur from the outset of the design and construction process to successfully build a sustainable infrastructure that achieves long-term benefits.
- Creating "sustainable" buildings is a natural extension of healthcare's mission to "first, do no harm." Healthcare organizations that pursue this mission and have committed to being world-class providers simply cannot ignore the profound, measurable effects that sustainable hospital environments have on healing.
- Designing and building greener facilities is a popular way for healthcare organizations to distinguish themselves from competitors in an increasingly competitive market – especially since the physical healing environment has been so closely linked to patient outcomes. Pursuing commonly recognized ratings such as LEED certification helps validate and publicize green building efforts.
- That a hospital is "green" does not mean that it functions less effectively than a conventional hospital, and also does not necessarily mean that it will cost more. Sustainable healthcare facilities look and, in many ways, function similarly to conventional healthcare facilities, but have the added benefits of conserving natural resources and showing concern for human comfort, indoor environment and employee productivity. Short-term increases in costs for building greener facilities are far outweighed by long-term savings from increased efficiency.

About the organizations

LEED is a Green Building Rating System developed by the US Green Building Council. LEED aims to recognize environmental leadership in the realty industry. Ratings are classified into four categories: LEED Certified, Silver, Gold and Platinum, with Platinum being the highest rating. In order to achieve LEED certification, buildings are rated in five key areas: sustainable site development, water savings, energy efficiency, materials selection and indoor environmental quality.

Kohinoor Hospital is the first hospital not only in India, but in Asia, to achieve Platinum rating. The 150-bed, multi-specialty hospital is located in Mumbai. The Certifying Authority for Kohinoor Hospital is the Indian Green Building Council (IGBC). The aim of IGBC is to assist in the creation of high-performance, healthful, durable, affordable and environmentally sound commercial and institutional buildings.

Dell Children's Medical Center of Central Texas, a member of the Seton Family of Hospitals, was the first hospital in the world to receive the LEED Platinum designation, given by the US Green Building Council. The 470,000-square-foot hospital opened in June 2007 on 32 acres that was once part of the former Mueller Municipal Airport, which has become one of the largest redevelopment efforts in the City of Austin's history.

Sources

Allen, J.G., MacNaughton, P., Laurent, J.G., Flanigan, S.S., Eitland, E.S. and Spengler, J.D. (2015). Green buildings and health. July 10. Available at www.ncbi.nlm.nih.gov/pmc/articles/PMC4513229/ (accessed August 6, 2017).

Dell Children's Hospital gets LEED platinum. Available at www.bdcnetwork.com/dell-childrens-hospital-gets-leed-platinum.

Dell Children's Medical Center (2008). Dell Children's Medical Center is world's first Platinum hospital. Available at www.dellchildrens.net/about_us/news/2009/01/08/dell_childrens_medical_center_is_worlds_first_platinum_hospital_2.

Express Healthcare (2011). Kohinoor is a LEEDing light. July. Available at www.expresshealthcare.in/201007/strategy02.shtml.

Huang, Y. (2011). *Impact of Green Building Design on Healthcare Occupants – With a Focus on Healthcare Staff* (unpublished Master's thesis). Michigan State University. Available at https://d.lib.msu.edu/etd/363/datastream/OBJ/download/IMPACT_OF_GREEN_BUILDING_DESIGN_ON_HEALTHCARE_OCCUPANTS___WITH_A_FOCUS_ON_HEALTHCARE_STAFF.pdf (accessed August 6, 2017).

Kohinoor Hospital – Platinum LEED certified green hospital sets benchmarks. Available at www.hospitalinfrabiz.com/kohinoor-hospital-ndash-platinum-leed-certified-green-hospital-sets-benchmarks.html.

Practice Greenhealth (2012, September 13). [Past Event] – Green Operations Series: Engaging employees & marketing the sustainable mission. September 13. Available at https://practicegreenhealth.org/webinars/green-operations-series-engaging-employees-marketing-sustainable-mission (accessed August 6, 2017).

Vittori, G., Geunther, R. and Glazer, B. (2013). Study: Extra costs minimal for LEED-certified hospitals. October 22. Available at www.usgbc.org/articles/study-extra-costs-minimal-led-certified-hospitals (accessed August 6, 2017).

Wang, H. and Horton, R. (2015). Tackling climate change: the greatest opportunity for global health. June 22. Available at www.thelancet.com/journals/lancet/article/PIIS0140-6736(15)60931-X/abstract.

Leadership imperative

One of the principal strategies of leadership in sustainable healthcare organizations is helping colleagues understand why sustainability is core to health and how sustainability is translated into the organization. One effective way of branding the organization with sustainability values is by investing in facilities that communicate such values. The visible commitment to sustainability explicitly promotes the health of patients and employees. It is the leader's responsibility to challenge the design and construction team to collaborate with the brand team – an unlikely team in years past.

A leadership dialogue

1 How can leaders use the process of building design to communicate sustainability values and strengthen brand recognition?
2 Why is the LEED for Healthcare rating system considered a guide for leadership strategic planning?
3 How is the relationship between employee recruitment and sustainability values reflected within the hospital's built environment?
4 What are the components of leadership's business case for meeting LEED standards?
5 How can leaders make the case for considering green leasing?

Syllabus integration

Overview

Students will learn from a healthcare design professional guest lecturer, who will explain the LEED rating system, exploring how to balance healthcare capital improvement demands with environmental challenges. Students will be exposed to resource and code constraints, programmatic requirements and institutional cultural barriers. Students will also explore the relationship between human health and the built environment.

Objectives

1 To make the connection between human health and the built environment.
2 To make the business case for sustainable building design.
3 To measure, manage and communicate sustainable design value to the community.
4 To explain how leveraging sustainable design can help create value by enhancing brand, reputation, innovation and leadership.
5 To visit a sustainably designed healthcare organization to assess sustainable design.

Topics

- WELL Building Standard
  Transparency tool (Google, Healthy Building Network and Perkins+Will)
- LEED: a LEED exercise with an architect/design professional
- The Living Building Challenge

Homework

Houghton, A., Vittori, G. and Guenther, R. (2009). Demystifying first-cost green building premiums in healthcare. *Health Environments Research & Design Journal* 2(4): 10–45.

IWBI (2017). WELL Building Standard. Available at:www.wellcertified.com/standard.

Kresge Foundation (n.d.). *How Do I Build Green?* Troy, MI: The Kresge Foundation, Green Building Initiative.

Transparency Tool. Available at http://transparency.perkinswill.com/.

Additional resources

Practice Greenhealth and the Institute for Innovation in Large Organizations (2008). The business case for greening the healthcare sector.

Practice Greenhealth (2009). Design & construction series: Metro Health Hospital's transition to a LEED hospital: design, construction and operations. Webinar, August 2014. Available at www.practicegreenhealth.org.

Notes

1 US EPA (2010). *Green Building*. Washington, DC: Environmental Protection Agency. Available at www.epa.gov/greenbuilding/pubs/about.htm.

2 Health Resources and Services Administration (n.d.). Hill–Burton free and reduced-cost health care. US Department of Health and Human Services. Available at www.hrsa.gov/gethealthcare/affordable/hillburton/.

3 Anon. (n.d.). A comparison of the world's various green rating systems. *FM Link*. Available at www.fmlink.com/article.cgi?type=Magazine&title=A%20comparison%20of%20the%20world%27s%20various%20green%20rating%20systems&pub=RFP%20Office%20Space&id=31124&mode=source.

4 Northbridge Environmental Management Consultants (2003). *Analyzing the Cost of Obtaining LEED Certification*. Westford, MA: Northbridge for the American Chemistry Council. Available at https://greenbuildingsolutions.org/wp-content/uploads/2016/05/LEED-Cost-Analysis-Report.pdf.

5 Ibid.

6 Ibid.

7 Gow, P. (2012). *An Experience of "Yes": Independent Schools Begin to Explore and Exploit the Power of Design Thinking*. Washington, DC: National Association of Independent Schools. Available at www.nais.org/magazine/independent-school/spring-2012/an-experience-of-yes/.

8 Shontell, A. (2011). 11 unusual ways Steve Jobs made Apple the world's most admired tech company. *Business Insider*, October 6. www.businessinsider.com/steve-jobs-apple.

9 Ibid.

10 White, G. (2009). Introduction to green leasing. *Green Real Estate Law Journal*, February 6. Available at www.greenrealestatelaw.com/2009/02/introduction-to-green-leasing.

11 Healthier Hospitals Initiative (n.d.). Advocate health care. Available at http://healthierhospitals.org/get-inspired/leadership-spotlight/advocate-health-care.

12 Fiksel, J., McDaniel, J. and Mendenhall, C. (1999). *Measuring Progress towards Sustainability Principles, Process, and Best Practices*. Columbus, OH: Batelle Memorial Institute. Available at www.eco-nomics.com/images/Sustainability%20Measurement%20GIN.pdf.

11

WATER CHANGES EVERYTHING

Community benefit + water management

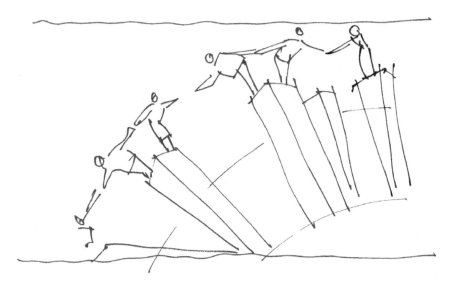

FIGURE 11.1 Growing community

Executive summary

Hospitals are reaching for a new definition of their role in community health that goes beyond traditional measures. Hospitals are uniquely posed in terms of one of mission, intellectual resources and reach into centers of influence. Hospitals have the ability to command attention and the leadership potential to help guide the community toward improved sustainability. This is the full implication of sustainability for hospitals and the basis of their new social compact. This new social compact includes a responsibility to ensure that potable water remains potable. One of the basic public health imperatives is access to clean water for drinking, bathing, food preparation and agriculture, among other daily activities.[1] Yet less than 1 percent of the Earth's water is potable.[2] Healthcare leaders have a role to play in ensuring that their organizations are not contributing to water contamination and are, instead, supporting efforts to protect the public's natural health resources.

A told story

DR. JENNY BEA: [*Pediatrician*] Welcome to the Memorial Hospital medical staff meeting, Dr. Goldberg. I had multiple patients come in last week complaining of the drizzlies and skin infections. The common thread between them is that they all swam in the river on Mellen Street for summer camp and said they may have swallowed water. I contacted the Health Department and discovered that the river water is contaminated.

DR. MORDECHAI GOLDBERG: [*Family physician*] Contaminated water? Where is it coming from? Not us, I hope. It's interesting you ask, because I asked the same question. I'm not sure where the water contamination is coming from, but I'm in the process of following up with the Health Department and will be sure to share with you what I learn.

DR. BEA: What I know is that I'm regularly visited by parents who come in dragging their kids along – they just want their kid's earache fixed. They don't care about overuse of antibiotics so long as the earache goes away. I spend a fair amount of time giving instructions about how to take the antibiotics, but no time on how to properly dispose of the excess drugs, if there are any. Before now, that step of the process simply hadn't crossed my mind. I focused on alleviating the child's pain.

DR. GOLDBERG: I can't say I've given water contamination much thought either until this point. For all I know, my patients could be dumping unused drugs down the toilet, contaminating the water supply and not realizing they are contributing to the problem. Sounds like we need to be thinking about clean water as an end in itself, rather than as a means to an end.

DR. BEA: I agree. [*Dr. Grisha Vladimir, the curmudgeonly orthopedic surgeon, joins the conversation along with Dr. Shmuel Cohen, the young hospitalist who is new to Memorial Hospital.*]

DR. GRISHA VLADIMIR: All I know is that they closed my favorite trout stream due to the water contamination. Now I can't take my grandkids fishing.

DR. SHMUEL COHEN: Here we are talking about water supply out in the community, while we can't even get it right in our own facility. [*He lifts a plastic water bottle from the table and places it next to his reusable bottle.*] We keep serving water in plastic bottles. Don't people realize that the water from the faucet is safe to drink, free and produces far less waste? We're privileged to have that luxury, compared with other places in the world, and yet we still go out and buy bottled water. It's so wasteful.

Healthcare systems are among a community's largest consumers of water. Consumption, however, varies greatly: Water use per capita ranges from 40 gallons per day to 350 gallons per day, depending on such factors as geographical location; services provided; size, age, and type of buildings; and water-use equipment and practices. A study of seven Massachusetts facilities found the potential for a reduction of almost 20 percent. Limiting water consumption provides hospitals with savings related to water supply and sewer costs. Some measures also can save energy costs, such as those that reduce the need for hot water. Water conservation also reduces the energy needed off-site to transport and process water and enables hospitals to be proactive stewards of their community's natural resources.

(US Department of Energy, 2011)

Stop dumping in the river

Hospitals have a unique opportunity to provide community leadership. It is a complicated process to convert that commitment into clear and achievable health objectives for a specific community. The process requires informed, thoughtful and committed leaders to focus institutional attention on the consequences of unsustainable systems.

Not-for-profit healthcare organizations have a community charter. The hospital receives tax benefits in exchange for promoting the community's health, including the provision of charitable care and other community benefits. For-profit hospitals are not tax exempt, but still typically provide some quantity of charitable care.[3]

NOT-FOR-PROFIT

The vast majority of hospitals in North America are organized as private, not-for-profit institutions. In Canada, nearly 70 percent of healthcare spending comes from public funding (Allin and Rudoler, 2016). In the United States, around 10 percent of all hospitals are for-profit institutions, whereas 70 percent are not-for-profit hospitals and the balance is public institutions (generally run by counties or municipalities).[4]

While there is great variation, many European nations offer publicly funded health services.[5]

It is the healthcare leader's professional obligation to practice sustainable behavior as a benefit to the community, regardless of the hospital's for-profit or not-for-profit tax status. When leadership's business decisions benefit the operators and the people inside the system at the expense of innocent third parties (i.e., community members) leadership is not fulfilling its professional obligation. Said differently, when the hospital pollutes potable water sources, it is at the expense of the community and contradicts the healthcare leader's professional imperative.

Pharmaceutical waste dumping in potable water sources is a real concern (Conley, 2011; WHO, 1999; Owens, 2015). Both China and India are larger producers of antibiotics and suffer from poor water quality regulations (Litovsky, 2016). In addition, there is evidence of antibiotic-resistant bacteria outflowing and breeding from waste water treatment plants. This is cause for concern, as these resistant genes have the opportunity to spread (Olena, 2013). The Joint Commission has guidelines on pharmaceutical waste, as does the EPA through the Resource Conversation and Recovery Act (RCRA) since 1976. RCRA regulations were written for the commercial industry and did not consider the potential impact of hospital-based pharmaceuticals. The US EPA audits hospitals with fines of up to $32,500 a day; hospitals audited have received fines of between $40,000 and $250,000.[6] "Recent concerns regarding the documentation of drugs in drinking, ground, and surface waters have led to a rapid rise in public awareness and calls for action at the federal, state, and local levels."[7]

Are the implications of healthcare sustainability reflected in community benefit calculations? Many healthcare organizations are content with their traditional roles, equating community benefit activities with being responsive to tax-exemption requirements around indigent care. It is an immensely important contribution, but a level of engagement that often falls short of the organization's broader potential contribution. By the very definition of the healthcare mission, it may be argued that the achievement of sustainability requires a broader engagement by hospitals. Accountability is a timely topic owing to not-for-profit healthcare's focus on the Internal Revenue Service (IRS) 990 form. Could sustainability be included as part of the IRS 990 forms? Yes, hospitals have the opportunity to enhance community benefit through sustainability. The question is not: how is sustainability related to reporting? The question is: to what extent is sustainability a fundamental dimension of the value of community benefit and accountability to the community?

WATER FACTS

Over 70 percent of the Earth's surface is covered with water. Approximately 97 percent of the Earth's water can be found in oceans.[8]

WHERE IS THE WATER ON A HOSPITAL CAMPUS?

- Cooling towers
- Faucets
- Flow controls
- Ice machines
- Irrigation
- Kitchens

- Laundry
- Showers and tubs
- Sterilizers
- Toilets
- Vacuum systems
- Washers
- Water treatment

ALTERNATIVE SOURCES OF WATER SOURCES/WAYS TO GAIN ACCESS TO WATER

- Air-conditioner condensate
- Cooling tower
- Foundation drain water
- Gray water
- On-site treated waste water systems
- Rainwater harvesting
- Stormwater harvesting

Water is a zero-sum game

Potable water is a finite resource, of which healthcare organizations are major consumers. To put it explicitly, the more water the hospital uses, the fewer resources the hospital affords the rest of the community. It is natural for healthcare leaders to be concerned about the community's water supply when considering the 24-hour-per-day, seven-days-a-week, 365-days-a-year nature of healthcare operations.

FIGURE 11.2 Water, water everywhere

The imperative

- Vital water resource management.
- To completely stop leakages from taps and other pipes within the hospital with the aim of conserving water and reducing pollution.
- To educate staff about the importance of water conservation, ultimately strengthening staff resolve on the issue.

The initiative

Water consumption at Victoria Hospital was observed to be high, standing at 417 liters/bed/day. This figure raised concern among senior management. Once the hospital became a member of the Global Green and Healthy Hospital (GGHH) initiative, the hospital embarked upon reducing its water usage by repairing leaking pipes and taps within the hospital. The leaking pipes within the hospital were identified and suitable replacements were installed. The initial amount of water consumed every month was determined from a baseline assessment. After the leaking pipes were corrected, there was a significant improvement in the amount of water saved.

The sustainability strategy implemented started with an investigation which identified areas within the hospital that had leaking taps. This was illustrated in a map so that appropriate measures could be implemented, which helped track and monitor water use. Hospital staff were educated about situations to avoid so as not to waste water. Staff were proactively trained about how to prevent excessive water use for their activities. The task team, comprising cleaners and housekeepers, helped ensure that taps and other water sources were regularly turned off when not in use. A framework that aspires to net zero water use was developed with 32 cleaners and 12 housekeepers who were appointed to lead the project. This group was chosen because they play frontline leadership roles in maintaining hospital hygiene. They were trained by the management to report broken and leaking taps. Hospital staff changed their practices and stopped wasting water.

A training workshop was conducted for the task team charged with this responsibility. They were trained about how to identify leaking pipes and other water conservation strategies. Furthermore, a plumbing expert installed efficient faucets and toilets that use less water, and fixed leaking pipes. A system was put in place where there is routine monitoring of plumbing and pipes to prevent leaking. Training was conducted by the CEO and Support Service Manager. During the training the cleaners and housekeepers were given authority to switch off unnecessary water and report wastage to the Support Service Manager. The hospital also bought a new autoclave machine which saves 200 liters of water per cycle, which contributed to both cost and water savings.

The metrics

Tracking Progress Expenditure reports were used to track progress. Before sustainability efforts commenced, Victoria Hospital spent an average of R30,000 per month on water. Six months after the water use reduction project was implemented, the hospital was paying an average of R23,000 per month. A monthly meeting was established to gather feedback sessions and to discuss challenges that the cleaners and housekeepers experienced. While not well measured, it was observed that the attitude and mindset of people when using water changed as people were keen to reduce the hospital water costs.

Lessons learned

At the beginning of the project, there was limited staff interest because people did not understand the need for water conservation. Only after we conducted a community awareness campaign about water conservation did we achieve buy-in, especially from nursing and kitchen staff who ultimately showed keen interest. Behavioral change was another challenge; it was difficult for people to stop bad practices such as leaving taps running while brushing teeth and washing dishes. We educated colleagues about water conservation using illustrations to explain the amount of water used doing those activities and how much the hospital is paying for a liter of water. This visual explanation provided a clear picture about the impacts of individual behaviors. It is not easy to change behavior. The task team was persistent in their activities, which were instrumental to achieving water conservation goals. For our next steps we aim to use food waste from the kitchen to develop a food garden within the hospital. This program will help reduce food waste while creating a source of fresh oxygen and clean air from the plants in the garden.

About the organization

Victoria Hospital is a secondary district hospital situated in Wynberg, a southern suburb of the Western Cape, South Africa. It is a 158-bed hospital known for its high standard of care as well as its concern for the population it serves. Victoria Hospital provides a range of both medical and surgical services. The hospital building itself is about 11,985m^2 sitting on land of about 12,125m^2.

Sources

This content has been adapted from Global Green and Healthy Hospitals Initiative Case Studies, with permission.

There are many visible, if sometimes subtle, cues that patients and fellow clinicians identify regarding water management practices. For instance, the patient may see the nurse leave the water running after she washes her hands, or the physician may drink bottled water from the vending machine rather than carry a reusable water bottle. A patient who cares about sustainability may conclude that neither her clinical team nor the hospital cares about water stewardship.

COST OF WATER

Water is one of the most precious resources, not only for its importance as a vital human necessity, but also because of the associated financial value. According to the Earth Policy Institute, water prices have increased over the past five years:[9]

- 27 percent in the United States
- 32 percent in the United Kingdom
- 45 percent in Australia
- 50 percent in South Africa
- 58 percent in Canada
- In Tunisia, the price of irrigation water increased fourfold over a decade.

For the average use of water for a resident of Washington, DC, 127,400 gallons at the price of $380, a vendor in the slums of Guatemala would charge more than $1,700.[10]

HOW MUCH DOES EACH DROP COST?

In addition to using more than 2.7 million tons of petroleum-based plastics every year, the cost of water can be as much as a gallon of gas. "That 9-ounce bottle of Evian spring water at $1.49? That'll cost you about $21 a gallon." That cost can be up to 1,900 times more than tap water (Scheer and Owens, 2011).[11]

FRED: [*Fred realizes he is drinking from a plastic water bottle he purchased in the vending machine earlier that day.*] What if we eliminated bottled water to support our sustainability effort? Don't look at my bottled water, please. I forgot my reusable bottle at home today.

AMALIE DUBOIS: [*Sustainability Director*] I like the idea of eliminating plastic bottles, though we need to be cognizant that bottled water sales generate revenue for the hospital. Eliminating bottled water would certainly mean taking a stance that is unlikely to be fully appreciated, though it's the right thing to do.

FRED: What if we simply ban plastic water bottles? We don't provide, don't stock and don't process purchase orders for bottled water? I'm talking about putting a ban on bottled water for meetings. We could serve water from pitchers, right?

AMALIE DUBOIS: Right.

FRED: This decision will be a politically charged issue. It's important not to eliminate bottled water, only to turn around and spend a million dollars installing water purifiers on every tap water outlet. Just in case we're questioned, how does the quality and safety of our tap water measure up?

AMALIE DUBOIS: Our region boasts one of the top-ten purified waters in the country. The water coming out of the sink faucet is tested 300 to 400 times a day in this area.[12] There's nobody regulating the quality of bottled water, other than studies that show there's more bacteria leaching from plastic water bottles than exist in the tap water.[13]

FRED: Really?

AMALIE DUBOIS: Yes. I know Dr. Goldberg and Dr. Bea expressed interest in the topic of water management as it relates to the public's health because they both recently had ill patients as the result of contaminated water. Maybe they'll help us champion water management among the physician community. We could give reusable water bottles as a gift during Doctor's Day. Carrying a reusable water bottle is a visible way of demonstrating commitment to sustainability. [*Dr. Hedge, Chief Medical Officer, enters the room.*]

DR. HEDGE: Good afternoon, Amalie. I've been thinking about our hand-washing policy and reflecting on this morning's medical staff meeting. I'm sure you heard that there was dialogue about our responsibilities with regard to water stewardship. From a sustainability perspective, do you recommend that we build our policy around hand washing or around foam-in-foam-out? I'm guessing that foam-in-foam-out might conserve more water. Clinically, there's no proven difference in safety outcomes.[14]

AMALIE DUBOIS: Using water for hand washing has less of an environmental impact than using foam.[15]

DR. HEDGE: It does? That's counterintuitive.

AMALIE DUBOIS: I was surprised to learn that fact too. But think about all the containers that the foam is packaged in, all of the transportation that is involved – none of that reflects sustainable values. Besides, there are usually chemicals in the foam that aren't healthy when used over an extended time. In fact, the FDA [*US Food and Drug Administration*] has raised concerns about some of the chemicals used in foam sanitizer products.[16]

FRED: Sounds like our water management policy is starting to take form. Amalie, it'll be your responsibility to coordinate with Dr. Hedge to ensure all our physicians receive a reusable Memorial Hospital water bottle for Doctor's Day. Also, please coordinate with Dr. Hedge for our hand-washing policy. Let's underscore our unrelenting commitment to clinical quality and safety with an appreciation for sustainability implications.

DROP BY DROP

WaterSense is a US EPA program that labels water conserving-related consumer goods such as bathroom sinks, toilets, flushing urinals and showerheads. Americans use over 550 liters of water per day, while their Ugandan peers use less than 25 liters per day (UNDP, 2010).

WATER = ENERGY

Both water and energy are fundamental natural resources that are intertwined on many healthcare campuses. Energy is required to store, treat, transport as well as heat and cool water. Conversely, a significant amount of water is used in cooling and heating towers to produce energy, such as the hospital cooling tower.[17]

Leadership imperative

Most not-for-profit healthcare organizations are content with their traditional roles, equating community benefit activities as defined by the IRS through which they have obtained their tax exemption. It is an immensely important contribution, but at a level of engagement that often falls short of the organization's potential contribution. To what extent is sustainability a fundamental dimension of the value of community benefit and accountability to the community? Should the value of sustainability net cost contributions be included on IRS form 990? It is likely that future regulators will, in some fashion, recognize the value of creating clean air, clean water, reduced waste disposal and improving community health as much as indigent care and medical education. This is the future of health that may well reach even the IRS.

The unintended consequences of well-intended actions can have sustainability implications, such as in the case of using foam for hand washing. Sustainability practices may indirectly impact patient care. Although the relationship between water management and community benefit can be challenging, the reality is that hospital leaders have significant opportunities to positively impact the community by managing water processes according to sustainability principles. The most direct opportunities are through organizational operations, such as disposing of pharmaceutical waste according to Joint Commission and EPA standards, and by reinforcing the actions that individuals can take to be effective water stewards, such as drinking from a reusable water bottle and hand washing.

FRED: Community benefit is why we exist. Everything we do ought to be about community benefit, regardless of whether or not we get IRS credit for it.

CHAIRWOMAN OF THE BOARD: Very true. Very true.

FRED: It's interesting; up until this point I really hadn't thought about how we might incorporate sustainability into our community benefit framework.

CHAIRWOMAN OF THE BOARD: You know, Fred, I think it's about time we rethink our views on community benefit. Sure, we're a nonprofit organization and community benefit is the mission we were founded on. But what are the full implications of our community benefit, and how does sustainability weigh in?

Leadership dialogue

1   How can healthcare leaders lead by example in response to the increasingly important and visible community issue of water management?
2   What is the validity of the perception that a leader using a reusable water bottle is a display of "political correctness"?
3   Realizing that the elimination of bottled water from the hospital may be unpopular with some staff members, what are the pros and cons of eliminating bottled water to signal the institution's commitment to improved water management?
4   Water is used to create energy, and energy is used to process water. Why does or doesn't linking water and energy management engage leadership in sustainability strategies?
5   How could administrative leadership engage physicians to help lead sustainability improvements?

Practical application: the important role of water

The imperative

•   Reducing risk from long-term water scarcity and rising water prices.
•   Protecting health by increasing access to clean drinking water.
•   Reducing impacts of healthcare upon water pollution levels.
•   Achieving financial savings through water and energy efficiency.
•   Engaging employees in a sustainability program.
•   Acting as a responsible community member.

The initiative

Water fills different social and environmental roles throughout the world. The approach of healthcare organizations to water management in developed regions is often one of conservation and preservation – reducing water use and eliminating pollutants when possible. Such measures often target operations and equipment, sanitary and laundry fixtures, building maintenance, food service practices and outdoor water use, and include actions such as:

•   Installing low-flow fixtures and irrigation systems.
•   Retrofitting or repairing inefficient machinery.
•   Promoting closed-loop heating, ventilation and air-conditioning systems that eliminate once-through water use.
•   Collecting rainwater and gray water from other sources for non-potable use.
•   Managing water pollution through use of technologies such as permeable pavers, bioswales and rain gardens.

The concept of water management embodies a different meaning in less developed locations that face issues such as lack of reliable water infrastructure, lack of access to clean drinking water and overall water scarcity. Hospitals in such regions develop innovative solutions to manage common water-related issues within both their organizations and their local communities. Examples of such solutions include the following:

- Gaviotas Hospital in Colombia uses a deep-soil water pump and simple solar energy distillation technology to provide purified water to the hospital and local community members.
- Kisii Hospital in Kenya partners with international organizations to install new water pumps within a local community. Water for the hospital is stored and delivered overnight, leaving the pump free for community access throughout the day.
- Sambhavna Clinic in India captures rainwater for washing laundry and irrigating its on-site garden of medicinal plants with naturally treated waste water from its facilities.

The metrics

Regardless of the external water availability context, healthcare organizations throughout the world measure water in the same ways:

- Quantity, cost and types of water used.
- Savings from water conservation and efficiency projects.

Lessons learned

- Making employees part of the process – whether asking them to report leaks, audit water fixtures or suggest ideas for water conservation – brings additional perspectives to the table and contributes to the long-term success of a water management program.
- Water is not just a behind-the-scenes operational issue – it can also be a mechanism to engage the community and employees through community classes and demonstrations about conservation measures, such as water-efficient landscaping, making a rain barrel or building a rain garden.
- Despite the differences between hospitals around the world, many hospitals have similar basic water uses in equipment, operations and procedures – uses such as hand washing, laundry, sterilization and water-using medical equipment. Globally, the vast range in water usage between facilities of similar size and scope indicates that most hospitals are using much more water than is actually necessary. Hospitals in regions where the water supply is abundant can learn conservation tactics from those organizations where water is less easily accessible.

About the organizations

Gaviotas Hospital is located in a small, self-sustaining town of 200 people in the remote plains of Colombia. The 16-bed rural hospital facility was self-sufficient, running entirely on natural processes and renewable energy sources until the building was closed in the early 1990s and converted into a bottling plant that now provides free drinking water to local residents.

Kisii Hospital is a Level 5 regional referral hospital that covers the South Nyanza, South Rift and entire Gusii Regions of Kenya. The 450-bed hospital offers basic services in curative, promotive, preventive and rehabilitative care.

The Sambhavna Clinic in Bhopal, India offers free medical care and rehabilitation through allopathy, ayurveda and yoga to individuals affected by the 1984 Bhopal gas disaster. It is located on about an acre of medicinal herb garden in the heart of the gas-affected area of Bhopal.

Sources

Allin, S. and Rudoler, D. (2016). International health care system profiles. Available at http://international.commonwealthfund.org/countries/canada/ (accessed August 4, 2017).

Anon. (2007). Las Gaviotas: sustainability in the tropics. *World Watch Magazine* 20(3). Available at www.worldwatch.org/node/5020.

Ellis, J. (2006). 'Gaviotas.' *WorldChanging*, September 10. Available at www.world changing.com/archives/004910.html.

Fay, A. (n.d.). Kenya – Kisii Hospital Project Report. *Blue Planet Network*. Available at http://blueplanetnetwork.org/projects/kenya.

Litovsky, A. (2016). Antibiotic waste is polluting India and China's rivers; big pharma must act. *Guardian*, October 25. Available at www.theguardian.com/sustainable-business/2016/oct/25/antibiotic-waste-pollution-india-china-rivers-big-pharma-superbugs-resistance (accessed August 4, 2017).

North Carolina Department of Environment and Natural Resources (2002). *Water Conservation Checklist*, Available at http://infohouse.p2ric.org/ref/23/22006.pdf.

Olena, A. (2013). Resistant wastewater. *The Scientist*, December 18. Available at www.the-scientist.com/?articles.view/articleNo/38730/title/Resistant-Wastewater/ (accessed August 6, 2017).

Owens, B. (2015). Pharmaceuticals in the environment: a growing problem. *Pharmaceutical Journal*, February 19. Available at www.pharmaceutical-journal.com/news-and-analysis/features/pharmaceuticals-in-the-environment-a-growing-problem/20067898.article (accessed August 4, 2017).

Scheer, R. and Moss, D. (2011). Drinking from a bottle instead of the tap just doesn't hold water. Available at www.scientificamerican.com/article/bottled-water-ridiculous-waste/ (accessed August 6, 2017).

Somnath Baidya Roy (2006). Making life possible. *Yes! Magazine*, August 31. Available at www.yesmagazine.org/issues/health-care-for-all/making-life-possible.

The Bhopal Medical Appeal (n.d.). The clinics. Available at www.bhopal.org/the-clinics. Official web portal for Kisii Level5 Hospital available at www.health careinkenya.com/696-kisii-level-5-hospital.

United Nations Development Program (UNDP) (Ed.) (2010). Average water use per person per day. April 15. Available at www.data360.org/dsg.aspx?Data_Set_Group_Id=757 (accessed August 6, 2017).

Syllabus integration

Overview

This lesson provides an overview of approaches that may be used to protect the quality of water that is filtered through healthcare systems.

Objectives

1   Describe the environmental significance of water quality data
2   Identify opportunities to efficiently manage water resources
3   Assess why and how the community benefits from water management
4   Define the link between human health, environmental sustainability and water quality

Topics

1   Factors that influence water operations
2   Relationship between water quality and human health
3   Risk categories of water management
4   Gray water analysis case study for promoting resource management, stewardship and conservation
5   Social factors that influence the adaptability of sustainable water programs

Homework

GEMI (2007). *Collecting the Drops: A Water Sustainability Planner*. Global Environmental Management Initiative, Water Sustainability Work Group.

GGHC (2007). *Reducing Potable Water Use Technical Brief*. Available at www.gghc.org/documents/TechBriefs/GGHC_TechBrief_Reducing-Potable-Water.pdf (accessed October 22, 2012).

GGHC (2007). *Sustainable Site Design: Stormwater Mitigation Technical Brief*. Available at www.gghc.org/documents/TechBriefs/GGHC_TechBrief_SiteDesign-Stormwater.pdf (accessed October 22, 2012).

WHO (2017). *The Health and Environmental Linkages Initiative (HELI) Water, Health and Ecosystem Overview*. Available at www.who.int/heli/risks/water/water/en/.

WHO (2017). *Water, Sanitation and Hygiene in Health Care Facilities: Status in Low- and Middle-income Countries and Way Forward*. Available at http://apps.who.int/iris/bitstream/10665/154588/1/9789241508476_eng.pdf?ua=1.

Additional resources

EPA (2012). *Saving Water in Hospitals.* Available at www.epa.gov/sites/production/files/2017-01/documents/ws-commercial-factsheet-hospitals.pdf.

Practice Greenhealth (2008). Design and construction series: low flow water fixture selection for healthcare facilities. Webinar, July 11. Available at www.practice greenhealth.org.

Practice Greenhealth (2008). Operations series: water use in facilities: laundry, stormwater and green roof case studies. Webinar, July 25. Available at www. practicegreenhealth.org.

Notes

1 Linn, A. (2007). Corporations find business case for going green. MSNBC, April 18. Available at www.msnbc.msn.com/id/17969124/ns/business-going_green/t/corporations-find-business-case-going-green/#.T8RMcNVPujN.

2 AWWA (n.d.). How much of the Earth is covered with water? Fairfax County Water Authority, American Water Works Association. Available at www.fcwa.org/education/education.htm.

3 Federation of American Hospitals (2003). *FTC/DOJ Hearings on Health Care and Competitive Law and Policy Statement of the Federation of American Hospitals – Hospital's Non-Profit Status.* Federal Trade Commission. Available at www.ftc.gov/ogc/healthcare hearings/docs/030410fay.pdf.

4 Taylor, D. (2002). What price for-profit hospitals? *Canadian Medical Association Journal* 166(11): 1418–19. Available at www.cmaj.ca/content/166/11/1418. full.pdf+html.

5 www.theguardian.com/healthcare-network/2011/may/11/european-healthcare-services-belgium-france-germany-sweden.

6 Smith, C. (2007). *Managing Pharmaceutical Waste: A 10-step Blueprint for Healthcare Facilities.* Houston, TX: PharmEcology Associates, LLC. Available at https://noharm-global.org/documents/managing-pharmaceutical-waste-10-step-blueprint-health-care-facilities-united-states.

7 Ibid.

8 USGS (1984). How much water is there on, in, and above the Earth? US Geological Survey. Available at http://ga.water.usgs.gov/edu/earthhowmuch.html.

9 Clark, E.H. (2007). Water prices rising worldwide, Earth Policy Institute, Plan B updates, March 7. Available at www.earth-policy.org/index.php?/plan_b_updates/2007/update64.

10 Ibid.

11 Carpenter, M. (2006). The pumped-up price of water, *Pittsburgh Post-Gazette*, March 17. Available at www.post-gazette.com/stories/sectionfront/life/the-pumped-up-price-of-water-432952/.

12 City of Toronto (n.d.). Living in Toronto: frequently asked questions. Available at www.toronto.ca/water/faq.htm.

13 Anon. (2010). Bottled water contains more bacteria than tap water. *Telegraph*, May 25. Available at www.telegraph.co.uk/health/healthnews/7763038/Bottled-water-contains-more-bacteria-than-tap-water.html.

14 Roylance, F. (2007). Manufacturers may be the only ones cleaning up. *Los Angeles Times*, February 11. Available at http://articles.latimes.com/2007/feb/11/news/adna-germs11/2.

15 Anon. (2008). Fate of personal care products linked to environmental pollution and human health concerns. *Science Daily*. www.sciencedaily.com/releases/2008/05/0805 16100942.htm.

16 US FDA (2010). Triclosan: what consumers should know. US Food and Drug Administration. Available at www.fda.gov/forconsumers/consumerupdates/ucm205999.htm.

17 NNEC (n.d.). *The Energy–Water Nexus.* New York: Network for New Energy Choices. Available at www.newenergychoices.org/index.php?page=ew_intro&sd=ew.

12

ENVIRONMENTAL ETHICS IN HEALTHCARE MANAGEMENT

FIGURE 12.1   Planting sustainability

Executive summary

The purpose of this chapter is to help leaders understand the ethics surrounding sustainability in healthcare. The emphasis is on the "why" more than on the "how." This chapter describes what is at stake if leaders operate according to the traditional, unsustainable healthcare paradigm.

The mission of healthcare organizations argues for sustainable management to improve the triple-bottom-line goals of community health. It is the ethical imperative of leadership to serve the health of the community, which includes the health of the environment that influences human health.

A told story

FRED: As it turns out, the Chairwoman of the Board's nudge to become a sustainable organization wasn't such a bad idea after all. Sure, it was painful to admit that I didn't have all the answers. I remember the Chairwoman wanted me to focus on "going green" while I was doing all I could just to survive. As for corporate responsibility, I now realize that sustainability is a foundational element of every institution's missional commitment to its community, and that is especially true for hospitals, which are fundamentally institutions of health and healing.

Mission alignment

There is more to sustainability than environmentalism and the threat to human health if we maintain the status quo.[1] Sustainability leadership provides the platform for the emergence of an emphasis on improving health, not just healthcare. The concept of sustainability is at once philosophical and operational, ethical and practical. Truly sustainable organizations invest in products and services that are fiscally, socially and environmentally responsible, regardless of the industry within which they operate.[2] At its core, sustainability is a natural component of improving public health and well-being.[3] Given the growing body of research that demonstrates the direct impact of the environment upon human health,[4] it is vital to recognize sustainability as a distinct social good.

Sustainability is a shared value across a variety of human endeavors and industries. However, healthcare's stake in sustainability is unique. Sustainability has a special relationship with healthcare ethics, with the health of the community, and with the many roles of healthcare organizations within the communities they serve. Sustainability aligns with the mission of healthcare organizations. The common mission of healthcare organizations is primarily related to improving the health of individuals and the health of the community. Is the goal of sustainability part of this mission itself, or is it a strategy for achieving this mission? For healthcare in particular, it would be inconsistent from an ethical leadership perspective to be excellent at quality and ignorant of sustainability.[5] Values associated with sustainability are part of the healthcare leader's commitment to professionalism.

Healthcare as an industry cannot independently solve the problems facing our environment, but it can be a model of response, a source of expertise and a catalyst of action. Embracing sustainability presents an opportunity to redefine the field of healthcare, both internally, within the hospital itself, and in the eyes of external stakeholders. Public commitment to sustainability provides a meaningful platform from which to make contributions to the health of the community. Through sustainability, hospitals face opportunities to explicitly exercise their commitments to prevention and wellness, thereby strengthening the compact between hospitals and the communities they serve.

The next frontier of leadership

Imagine a new leader phoning her healthcare administration professor after being assigned a sustainability task at work. The new leader says, "Professor, I'm in my first job and you neglected to teach me about sustainability. Why didn't you ever address sustainability during all the time I was in school?"

Our young leader chose a career in healthcare administration, believing that such a career would not only challenge and reward her, but would also allow her to make a positive contribution to the health and well-being of others. As she advances in her career, our healthcare leader is exposed to ethical questions that make her uneasy. For instance, she worries that hazardous pharmaceutical waste could be inadvertently discarded alongside normal trash. She wishes more attention would be devoted to how the hospital can discard its trash with confidence that it will be handled properly. She wonders whether there are mechanisms within the hospital capable of resolving these potential conflicts in ways that benefit patient and community well-being.

Sustainability resources and solutions exist in increasing numbers of healthcare organizations. Yet, hospitals throughout the country do not sort their waste sustainably, adding unnecessarily to landfills while increasing hospital costs for the disposal of otherwise recyclable content.[6] This is just one common example of how sustainability plays into healthcare leadership thinking. What, if anything, can healthcare leaders do to resolve this conflict for their organizations? Why should healthcare leaders care about sustainability in the first place? Exploring these questions requires informed, thoughtful and committed leaders to focus institutional attention and its impacts upon the health of the community.

CHAIRWOMAN OF THE BOARD: Fred, your ethical commitment is to improve the health of your patients and the community that Memorial Hospital serves. That commitment must be reflected in our organizational actions.

FRED: I understand your point. I also worry about employee engagement, which is, at its core, an exchange of values; it's a promise between the healthcare leader and my employees. The promise is that the leader will optimize involvement, engagement and employee opportunities to make meaningful contributions in the workplace. If I promise employees that they will be engaged in improving health, but operate in a way that degrades environmental health, then I've misled my employees. Ethics are as much about actions as they are about moral intent.

The future of sustainability lies in wellness. Sustainability programs are evolving to move beyond the triple bottom line and health of the environment to a broader, comprehensive view of wellness. Leaders must build the bridge between sustainability and wellness by focusing on the complete health of the individual as the ultimate end goal. Healthcare leaders would do well to leverage existing sustainability successes that support the development of wellness platforms.

Community benefit

Community benefit is especially important to not-for-profit healthcare organizations because sustainability is perhaps the most visible measure of organizational commitment to mission. Healthcare organizations have a community charter, in effect. The hospital receives tax benefits in exchange for promoting the community's health. In exchange for tax exemption, not-for-profit hospitals provide charitable care. For-profit hospitals are not tax exempt, but still often provide some quantity of incidental charitable care.[7,8] The point is that, regardless of tax status, it is the healthcare leader's professional mission to practice ethical behavior. When leadership's business decisions benefit the operators and the people inside the system at the expense of innocent third parties such as patients, leadership is not fulfilling its ethical promise to the community. That is the definition of unethical behavior: taking advantage of someone who is not a party to the basic business exchange, but who is harmed by the unauthorized taking of their interest. Certainly, many healthcare leaders do not consider sustainability as part of their ethical decision-making framework.

Why is sustainability an ethical issue?

FRED: You know, the reason I got into healthcare in the first place was because I wanted to make a difference in the lives of others. I was, and still am, committed to the belief that being a healthcare leader provides a unique opportunity to impact the world in a meaningful way.

CHAIRWOMAN OF THE BOARD: That's good to hear.

FRED: As we develop our new leadership strategies, I'm starting to get some of my old idealism back. It's been a great learning experience to see how we can achieve business results with a holistic strategy based on driving out waste, saving resources, moving upstream to prevent problems, and looking downstream for true life-cycle costs and benefits. It's led to a much more effective mindset focused on prevention long term. Our new organizational strategy is dedicated to both improving health and avoiding unintended harm. That's why I got into this business in the first place.

First and foremost, sustainability is central to the concept and definition of healthcare, and therefore it is the business of healthcare. Sustainability is about identifying root causes and fully analyzing the downstream effects, improving processes to reduce waste and improving the health of the community. One cannot define the health of an individual or community holistically without considering the environment. Second, holistic leadership incorporates an ethic of sustainability that is central to the direction in which social and public policy is headed. Worldwide, capitated and bundled payments are an emerging strategy for reducing healthcare costs by reimbursing providers for clinically defined episodes of care.[9,10,11] Moving away from fee-for-service reimbursement will provide incentives to improve the health status of the population, reduce utilization and control costs. One could call it holistic leadership for a sustainable triple bottom line of healthcare.

The narrow value framework

If the value of the organization is restricted to curative care and serving as a doctor's workshop, then it has what may be defined as a narrow value framework. The weakness in the narrow value framework is that it does not align with the broad scope of the community's health needs. Restoring clinical health is certainly one dimension of an individual's health needs, but only one dimension.

Take obesity as an example. The hospital that embodies the narrow value framework may treat the obese patient through surgery and solve the weight problem in the short term. It is a wonderful outcome for the patient, who has struggled and receives an improved quality of life owing to the restorative services of the hospital and surgical team. This value has been so substantial, in fact, that hospitals are well loved for their ability to restore health, even as they have not been particularly effective or incentivized to move upstream and prevent many of the root causes of the illnesses they treat.

The broad value framework

Most hospitals take a narrow view of where their resources can be deployed and which of the community's needs can be addressed. This choice is in part an ethical issue. The community has a broad set of health-related needs, and the hospital boasts an aggregation of skills and technology at considerable expense to the community's resources.

FRED: We've done some great things for our patients over the years. But I'm beginning to believe that we may have underperformed some of our potential as well. As we work to accelerate our improvement in quality, service and finances via our sustainability engagement approach, it seems we could think more broadly about our community service equation. There may well be a broader value equation we should be pursuing. We might win over the entire City Council, not just the Mayor, if we took on a broader community commitment that includes sustainability. I've always thought that it's leadership's business to address the community's health needs […]

CHAIRWOMAN OF THE BOARD: I would hope so […]

FRED: But now, in addition, I believe that, just as leadership should make it our business to raise awareness and educate the community about obesity, so too should leadership raise understanding of the consequences that impact our community's health and the planet's health. I guess we are back to that triple bottom line again; anything else seems incomplete in terms of our mission.

Current state

Sustainability is a form of accountability that translates into maximizing the productive life cycle of investments in the healthcare system.[12] Increasingly, communities across the country and the world believe that healthcare leaders should be accountable for business decisions, including decisions that impact the environment.[13] Healthcare leaders may be held to an even higher standard of accountability than commercial industry leaders because of their commitment to the Hippocratic oath: first, do no harm.[14] Given these expectations, future healthcare leaders are an ideal audience for sustainability education.[15] Leaders who shy away from their sustainability responsibilities are effectively shunning evidence-based management best practices that demonstrate short- and long-term cost savings as well as improved environmental and population health management.

Future state

The World Health Organization and the American Public Health Association are examples of leading health entities that cite the health of one's environment as a major indicator of human health.[16] The National Institute for Occupational Safety and Health and the Occupational Safety and Health Association focus on the impact of human-made built environments on inhabitant health and workplace productivity.[17] Practice Greenhealth, the Healthier Hospitals Initiative and the American Hospital Association were among the first organizations to partner with leaders in the healthcare delivery setting to identify environmental sustainability as a key component of the healthcare agenda.[18]

A paradigm shift is underway in which the future of healthcare could look noticeably different from the systems in place today. From the United States to Colombia, from China to Australia, national healthcare reform is in process.[19] In each instance, the national healthcare model highlights concerns for both population health management and individual health accountability. Part of this transformational change acknowledges environmental health influences on human health and well-being.

Successful organizations outside of the healthcare industry are already embracing and practicing sustainable business methods.[20] Now that successful business leaders in other industries are realizing the benefits of sustainability as a leadership value, the time has arrived for healthcare leaders to embrace a holistic leadership ethic. This new leadership ethic integrates sustainability values into an empowering and broad engagement of the total community to create a future for health.

Leadership imperative

It is unlikely that many leaders would argue that sustainability trumps all other goals. The principles of sustainability, fundamentally, are about doing more good and less harm with less waste. For most healthcare leaders today, sustainability means improving health, creating as little harm in the process as possible, and continuing to provide quality services with what is likely to be a shrinking supply of resources. This mindset is likely to be at the heart of priorities which healthcare leaders will address in the future. For example, whether the objective concern is a better patient experience, less waste cost, less energy cost, community goodwill, brand equity, employee engagement, physician engagement, the process of managing each objective will likely be about creating more health as an organizational outcome with as few negative side-effects as possible. Achieving business results while creating a sustainable organization is an international business trend with infinite applications to healthcare sustainability.[21]

Leadership dialogue

1   What is the place of sustainability in a healthcare leader's professional ethics?
2   Sustainability constitutes a new domain for many healthcare leaders, particularly the underlying issues such as policy questions, implications for public health and the technical aspects of sustainability. How do leaders acknowledge the limits of their expertise while providing direction for change?
3   How can the ethics of sustainability be framed pragmatically to align with healthcare leaders' interests and priorities?
4   How can healthcare leadership manage the competition for resources between sustainability goals and other mission-related goals?
5   How can the healthcare leader remain true to her professional commitments while maintaining important relationships with business leaders who may not share her enthusiasm for trade-offs to achieve sustainability goals?

Syllabus integration

Overview

This chapter is based on the concept that decision-makers who are responsible for preserving the value of our healthcare investments should take steps to mitigate environmental risks posed by the healthcare industry. Responsible behavior regarding healthcare sustainability supports community benefit; conversely, failure to address sustainability issues leads to profound questions about the exercise of governance and leadership responsibilities. This class is structured to start a conversation about leadership's role in taking measurable, evidence-based approaches to healthcare sustainability.

Objectives

1   Explore corporate transparency as related to healthcare sustainability business practice.
2   Describe the critical linkage between leadership training and ethics:

    a   Present elements of a leadership approach for managing healthcare ethics training needs.

Topics

1 Ethics of leadership
2 Ethics of healthcare
3 Employee and community training

Homework

American College of Healthcare Executives Code of Ethics (amended 2016). Available at www.ache.org/abt_ache/code.cfm (accessed July 30, 2017).

Prepare for a final oral presentation and written content submission.

Read Environment and health: 8. Sustainable health care and emerging ethical responsibilities. Available at www.ncbi.nlm.nih.gov/pmc/articles/PMC80732/.

Additional resources

Bridgestar (2009). *How to Develop Yourself as a Nonprofit Leader*. Boston, MA: Bridgespan Group. Available at: www.bridgespan.org/insights/library/careers/how-to-develop-yourself-nonprofit-leader (accessed October 22, 2012).

US EPA (2009). *Green Team Checklist*. Energy Star, US Environmental Protection Agency.

Forward: evolution as epilogue

There is an inherent tension within any collaboration that exists on three different axes: the relationship between the collaborators, the evolution of the authors as they do the work, and the evolution of the ideas about which they are writing. It is fair to say that writing this book has brought about substantial change in each of these dimensions as the process progressed to a conclusion. The authors believe that the changes in each of these dimensions are instructive to other leaders who seek to grow professionally and bring about the kind of transformation of the healthcare system that only passionate, committed leaders can effect. In this epilogue, we seek to share our experience and to offer ourselves as a case study in how we grew through the process of collective thinking about leadership and the very human dimension of coming to be both better leaders and better persons.

It is hard to imagine three more different collaborators by background than a Canadian-born civil engineer, a health system CEO with considerable tenure in that role and a global social entrepreneur. The organizing concept which galvanized this unlikely partnership was the idea that leaders in the healthcare sector do not understand enough about the current practice of sustainability to integrate it effectively into their leadership practices and priorities. The notion was that providing information about sustainability tools and change principles would empower leaders to enhance their personal competencies and create both more effective leaders and better strategic priority lists. This reorientation would improve the healthcare sector and enable leaders to achieve their mission, goals and individual career visions.

As the project progressed, the authors realized that their initial goals were substantially less powerful than the model of leadership that had begun to emerge. This new model was a compelling integration of sustainability principles and leadership imperatives that were meeting with great success in various locales in healthcare. In fact, it became apparent that sustainability ideas were capable of leveraging the existing agendas of healthcare leaders around quality, safety, cost, service, people, growth and community that provided an exciting vision of the future. A leadership approach that incorporated a holistic concept of organizational success, driven by a focus on a triple bottom line of organizational success, community success and sustainability success, was the vision that could energize the entire healthcare community. This new theory of leadership based on the core principles of sustainability and serving our patients, our community and environmental survivability was the only concept that could survive as a long-term model of success.

The authors did not begin their journey with this outcome in mind. In fact, they believed that three core principles would be apparent in this work.

- Each individual author would bring knowledge and perspective which was a unique and a useful ingredient in creating a guide for leaders.
- The principal problem was a lack of knowledge on the part of healthcare leaders.
- The work of writing this book would represent a useful contribution to the field of healthcare management, but would have little impact upon the authors themselves.

All three of these principles ended up being both incorrect and incomplete. After completing this book, the authors have become convinced of the following conclusions.

- The authors came to see change in their collective efforts, much as the ingredients of flour, salt, yeast and water do not define bread – the product is uniquely different and more than the sum of its parts. It was, in fact, the learning that came from the authors' interactions that created value, rather than the contributions of the individual contributors. Much as yeast cannot bring about great wine without transforming itself, the basic character of the authors had to meld with the grapes and juice of other collaborators to understand the discoveries that came from their collaboration.
- The missing ingredient that leads to suboptimal leadership performance in healthcare is not a lack of knowledge, but rather the lack of a holistic view of leadership based on the central mission of healthcare, which is to improve health. Knowledge is the essential grist which leaders bring to their craft. But it is much more important that the context for leadership be informed by the fundamental goals and motivations of the constituents and the owners of most community hospitals.
- The authors found that they both grew and evolved as leaders through the process of articulating this new leadership concept. While there may well be authors who begin a book with a well-developed thesis that is simply recorded

for the benefit of others, in this case the central concept evolved substantially as the authors themselves grew to understand the implications of the holistic view of leadership that emerged.

The authors celebrate the development of this new theory of holistic sustainability leadership as different people with different theories of leadership. We believe that the future of health must be joined by those leaders who believe that a triple bottom line of patient-centric, community-centric and environment-centric will provide the foundation for a sustainable healthcare system. Our new-found calling is to be the architects and advocates of this new concept of holistic leadership, which we believe will improve the health of the citizens and the communities we serve.

It will also contribute to sustaining life on our planet and those who depend upon it. We invite all who agree with this goal to join the future of sustainable healthcare leadership.

Carrie R. Rich, J. Knox Singleton and Seema S. Wadhwa

Notes

1 Dentzer, S. (2011). Embarking on a new course: environmental health coverage. *Health Affairs* 30(5): 810.
2 IFC Sustainability (2012). Leadership in sustainability and corporate governance. International Finance Corporation. Available at www.ifc.org/sustainability.
3 MPHI View on Sustainable Society (2012). Sustainability. Medicine and Public Health Initiative. Available at www.mphi.net/?p=303&lang=en-us.
4 Boone, T. (2012). *Creating a Culture of Sustainability*. Chicago, IL: UIC Medical Center.
5 Singleton, J. (1992). Preserving the environment can bring hospitals savings, improved stature. Presented at Modern Healthcare Weekly Business News: Our Environment: A Healthcare Commitment, Hyatt Regency, Crystal City, Arlington, VA, March 8–9.
6 WWF (2011). Hospital waste factsheet. World Wildlife Fund. Available at www.wwfpak.org/factsheets_hwf.php.
7 Day, K. (2006). Hospital charity care is probed. *Washington Post*, September 13. Available at www.washingtonpost.com/wp-dyn/content/article/2006/09/12/AR2006091201409.html.
8 Federation of American Hospitals (2005). Future of hospital care: how will we pay the bill? A hospital payment symposium, Washington, DC, July 15. Available at www.fah.org/fahCMS/Documents/Future%20of%20Hospital%20Care/Future%20Hospital%20Payment%20conference%20transcript%2007.15.05.pdf.
9 CMS (2012). HCAHPS: Patients' Perspectives of Care Survey. US Department of Health and Human Services, Centers for Medicare & Medicaid Services. Available at www.cms.gov/HospitalQualityInits/30_HospitalHCAHPS.asp.
10 Cromwell, J., Dayhoff, D. and Thoumaian, A. (1997). Cost savings and physician responses to global bundled payments for Medicare heart bypass surgery – innovations in fee-for-service financing and delivery. *Healthcare Financing Review* 9(1): 41–57.
11 Satin, D. and Miles, J. (2009). Performance-based bundled payments: benefits and burdens of this pay-for-performance strategy. *Minnesota Medicine*, October. Available at www.minnesotamedicine.com/PastIssues/October2009/SpecialReportOct2009/tabid/3209/Default.aspx.
12 Freund, F. (2009). *Business Model Concepts in Corporate Sustainability Contexts*. Lüneburg, Germany: Leuphana Centre for Sustainability Management.

13 GEMI (2004). Environment: value to the investor. Global Environmental Management Initiative. Available at www.gemi.org/resources/GEMI%20Clear%20Advantage.pdf.

14 NIH (2010). Greek medicine. US National Library of Medicine, National Institutes of Health, Department of Health & Human Services. Available at www.nlm.nih.gov/hmd/greek/greek_oath.html.

15 Rich, C., Wadhwa, S. and Singleton, J. (2012). How to teach environmental sustainability in healthcare management education. In *Teaching Sustainability*. Allendale, MI: Grand Valley State University.

16 Tsai, T. (2010). Second chance for health reform in Columbia. *The Lancet* 375(9709): 109–10. Available at www.thelancet.com/journals/lancet/article/PIIS0140-6736(10)60033-5/fulltext.

17 Johnson, J. (2008). Rapporteur's report: services sector. *Journal of Safety Research* 39: 191–4. Available at www.cdc.gov/niosh/topics/PtD/pdfs/Johnson.pdf.

18 Umbdenstock, R. (2011). Hospital environmental sustainability: video intro. American Hospital Association. Available at www.hospitalsustainability.org.

19 Thomas (2011). China's health reforms progress. *International Insurance News*, March 11. Available at www.globalsurance.com/blog/chinas-healthcare-reforms-progress-324520.html.

20 Warde, A. (n.d.). Going green to stay in the black. University of California. Available at www.uci.edu/features/2011/02/feature_marriott_110214.php.

21 Colburn, M. (2011) Overcoming cynicism: Lord Hastings' call to action at Net Impact. *GreenBiz*, November 2. Available at www.greenbiz.com/blog/2011/11/02/overcoming-cynicism-lord-hastings-call-action-net-impact.

APPENDIX

Sample letter to hospital group purchasing organization requesting less toxic products

Minnesota Technical Assistance Program RESOURCE

Dear <insert name>:

As a healthcare provider deeply committed to public health, <insert your hospital's name> is re-examining the ways in which our operations affect our community. In particular, given the link between environmental quality and health, we are seeking to reduce the size of <insert your hospital's name>'s "environmental footprint."

<insert your hospital's name> is interested in procuring safe and cost-effective medical products whose manufacture does not produce mercury – a heavy metal – or persistent, bioaccumulative toxin (PBTs) emissions; which do not contain such materials; and which do not produce them when disposed or treated.

In June 1998, the American Hospital Association and the U.S. Environmental Protection Agency signed a Memorandum of Understanding with the goal of reducing the volume and toxicity of the medical waste stream. The partnership is now Hospitals for a Healthy Environment. Specific goals include reducing total volume of waste by 33 percent by 2005 and 50 percent by 2010; virtually eliminating mercury-containing waste by 2005; and minimizing other PBTs produced by the medical sector.

These goals are to be achieved through pollution-prevention measures, such as the purchase of non-toxic medical products. Accordingly, we are asking <insert your GPO name> to inform us how you can help us and other members who wish to comply with the Memorandum of Understanding. Specifically, given the serious health risks posed by mercury and dioxins – highly potent persistent bioaccumulative toxins produced when polyvinyl chloride (PVC) is manufactured and incinerated – we are seeking <insert your GPO name>'s responses to the following questions:

- Which products that <insert your hospital's name> purchases from <insert your GPO name> contain mercury or PVC?
- Does <insert your GPO name> offer versions of these products that do not contain these materials, and if so, what are they?
- If <insert your GPO name> does not currently offer non-toxic versions of these products, does it plan to do so in the future?
- Is <insert your GPO name> willing to allow members to purchase non-toxic products off-contract without incurring financial penalties in doing so?

We appreciate your response to this inquiry. All sectors of the healthcare industry must cooperate with one another if the significant benefits to public health envisioned by the Memorandum of Understanding are to be realized. The environmental consequences of medical product purchasing can no longer be ignored. We look forward to working with you to achieve these worthwhile goals.

Sincerely,

Developed by the Minnesota Center for Environmental Advocacy's Polyvinyl Chloride (PVC) Task Force and provided by MnTAP.

REFERENCES

AASHE (2010). *Higher Education Sustainability Staffing Survey* (Rep.). Denver: Association for the Advancement of Sustainability in Higher Education.

About Energy Star (2017). Available at www.energystar.gov/index.cfm?c=about.ab_index.

About Inova (2017). Available at www.inova.org/about-inova/index.jsp.

A comparison of the world's various green rating systems (2017). *FM Link*. Available at www.fmlink.com/article.cgi?type=Magazine&title=A%20comparison%20of%20the%20world%27s%20various%20green%20rating%20systems&pub=RFP%20Office%20Space&id=31124&mode=source.

Advisory Board (2017). CMS: US health care spending to reach nearly 20% of GDP by 2025. February 16. Available at www.advisory.com/daily-briefing/2017/02/16/spending-growth (accessed August 10, 2017).

Agency for Healthcare Research and Quality (2013). Quality and patient safety. January 10. Available at www.ahrq.gov/professionals/quality-patient-safety/index.html.

A history of lean manufacturing (2016). Available at www.strategosinc.com/just_in_time.htm.

A holistic approach to environmental public health (2011). *National Environmental Health Promotion Network*, September 20.

American Hospital Association (2017a). *Fast Facts on US Hospitals* (Rep.). Available at www.aha.org/research/rc/stat-studies/fast-facts.shtml.

American Hospital Association (2017b). Waste. Available at www.sustainabilityroadmap.org/topics/waste.shtml.

American Medical Association's Code of Medical Ethics (Rep.) (2017). Available at www.ama-assn.org/ama/pub/physician-resources/medical-ethics/code-medical-ethics.page.

Arizona State University (2008). Fate of personal care products linked to environmental pollution and human health concerns. *Science Daily*, May. Available at www.sciencedaily.com/releases/2008/05/080516100942.htm.

Baier, P. (2011). How leading firms make their sustainability reports stand out. *GreenBiz*, May.

Balanced Menus Initiative (2017). *Health Care Without Harm*. Available at https://noharm-uscanada.org/issues/us-canada/balanced-menus-initiative.

Bansal, P. and Roth, K. (2000). Why companies go green: a model of ecological responsiveness. *Academy of Management Journal* 43(4): 717–48.

Bay Area Dioxins Project (2003). *Why Are Hospitals Rethinking Regulated Medical Waste Management?* (Rep.). Association of Bay Area Governments. Available at www.abag.ca.gov/bayarea/dioxin/pilot_projs/MW_Background.pdf.

Beeson, S. (2006). *Practicing Excellence: Five Reasons Medical Groups and Hospitals Striving for Culture Change Must Get Physicians on Board.* Gulf Breeze, FL: Five Starter Publishing.

Berchicci, L. and Bodewes, W. (2005). Bridging environmental issues with new product development. *Business Strategy and the Environment* 14: 272–85.

Berry, L., Mirabito, A. and Baun, W. (2010). What's the hard return on employee wellness programs? *Harvard Business Review*, December. Available at http://hbr.org/2010/12/whats-the-hard-return-on-employee-wellness-programs/ar/1.

Block, R. (2017). Boomers: you need to rethink seeking full time jobs with Gen Xers. Available at www.workforce50.com/content/general_resources_workforce50.html.

Boone, T. (2012). *Creating a Culture of Sustainability.* Chicago, IL: UIC Medical Center.

Bottled water contains more bacteria than tap water (2010). *Telegraph*, May 25. Available at www.telegraph.co.uk/health/healthnews/7763038/Bottled-water-contains-more-bacteria-than-tap-water.html.

Breast Cancer Prevention Partners (2017). Bisphenol A. Available at www.bcpp.org/resource/bisphenol-a/.

Brown, D., Berko, P., Dedrick, P., Hilliard, B. and Pfleeger, J. (2010). Burgerville: sustainability and sourcing in a QSR supply chain. *Business Administration Faculty Publications and Presentations* (33).

Brown, J. (2010). *Benchmarking Sustainability in Health Care Awards* (Rep.). Reston, VA: Practice Greenhealth. Available at https://practicegreenhealth.org/tools-resources/sustainability-benchmark-report-0.

Buchbinder, S.H. and Shanks, N.C. (2012). *Introduction to Health Care Management.* Burlington, MA: Jones and Bartlett Learning.

Buechner, S. and Ellis, M. (2011). 3 ways to improve green business rankings. *GreenBiz*, November.

Burns, J. (2011). At long last […] pay for outcomes starts to replace pay for performance. *Managed Care Magazine*, September. Available at www.managedcaremag.com/archives/1109/1109.payforoutcomes.html.

Button, K. (2016). 20 staggering e-waste facts. *Earth 911*, February. Available at http://earth911.com/eco-tech/20-e-waste-facts/ (accessed August 10, 2017).

Campbell, T. and Campbell, C.T. (2008). The benefits of integrating nutrition into clinical medicine. *Israel Medical Association Journal* 10: 730–2.

Caper, P. (2009). Health care should be driven by mission, not money. *Physicians for a National Health Program*, December. Available at www.pnhp.org/news/2009/december/health-care-should-be-driven-by-mission-not-money.

Carpenter, M. (2006). The pumped-up price of water. *Pittsburgh Post-Gazette*, March 17. Available at www.post-gazette.com/stories/sectionfront/life/the-pumped-up-price-of-water-432952/.

Cars, trucks, air pollution and health (2011). *Nutramed.* Available at www.nutramed.com/environment/cars.htm.

Casey, J. (2007). *A Lean Enterprise Approach to Process Improvement in a Health Care Organization* (Master's thesis). Massachusetts Institute of Technology.

Catalysis (2017). Available at https://createvalue.org/networks/healthcare-value-network/.

Centers for Disease Control and Prevention (2017, August 29). Overweight & obesity. August 29. Available at www.cdc.gov/obesity/data/index.html.

Centers for Medicare and Medicaid Services (2014). *HCAHPS: Patients' Perspectives of Care Survey* (Rep.). Available at www.cms.gov/Medicare/Quality-Initiatives-Patient-Assessment-Instruments/HospitalQualityInits/HospitalHCAHPS.html.

City of Toronto (2017). Toronto water. Available at www.toronto.ca/water/faq.htm.

Clark, E.H. (2007). Water prices rising worldwide. *Earth Policy Institute*, March 7. Available at www.earth-policy.org/index.php?/plan_b_updates/2007/update64.

Climate Change and Responsibility: Shareholders Press Boards on Social and Environmental Risks (Rep.). (2011). London: Ernst & Young.

Colburn, M. (2011). Overcoming cynicism: Lord Hastings' call to action at Net Impact. *GreenBiz*, November.

Collins, A. (2011). Food, health and the environment: the role of the clinician. *FoodMed.org*. Available at www.foodmed.org/2011/presentations/E2-Collins.pdf.

Covey, S.R. (1989). *The Seven Habits of Highly Effective People*. New York: Simon & Schuster.

Cram, P., Bayman, L., Popescu, L., Vaughan-Sarrazin, M., Cai, X. and Rosenthal, G. (2010). Uncompensated care provided by for-profit, not-for-profit, and government owned hospitals. *BMC Health Services Research* 10(90): 1–13.

Cromwell, J., Dayhoff, D. and Thoumaian, A. (1997). Cost savings and physician responses to global bundled payments for Medicare heart bypass surgery: innovations in fee-for-service financing and delivery. *Healthcare Financing Review* 9(1): 41–57.

Davis, K., Schoen, C. and Stremikis, K. (2010). *Mirror, Mirror on the Wall. How the Performance of the US Health Care System Compares Internationally* (Rep.). Washington, DC: Commonwealth Fund.

Day, K. (2006). Hospital charity care is probed. *Washington Post*, September 13. Available at www.washingtonpost.com/wp-dyn/content/article/2006/09/12/AR2006091201409.html.

Denning, E. (2011). Exercise in the workplace: part of Surgeon General's vision for healthy and fit nation. *Corporate Fitness Works*, February 18. Available at http://corporatefitness works.com/exercise-in-the-workplace-part-of-surgeon-generals-vision-for-healthy-fit-nation/.

Dentzer, S. (2011). Embarking on a new course: environmental health coverage. *Health Affairs* 30(5): 810.

Derkson, D.J. and Whelan, E.M. (2009). *Closing the Health Care Workforce Gap: Reforming Federal Health Care Workforce Policies to Meet the Needs of the 21st Century* (Rep.). Washington, DC: Center for American Progress.

Doman, J.L. (2007). *Leveraging Lean Process Improvement Methodology to Promote Economic and Environmental Sustainability: Obstacles and Opportunities* (Master's thesis). Rochester Institute of Technology, New York.

Dresser, T. (2012). Hazardous waste generation & management. *U.S. Environmental Protection Agency*, May 22. Available at www.epa.gov/region7/education_resources/teachers/ehsstudy/ehs9.htm.

Drewnowski, A. and Darmon, N. (2005). Food choices and diet costs: an economic analysis. *Journal of Nutrition* 135(4): 900–4.

EERE Information Center (2011). *Commercial Building Energy Alliances: Making the Business Case for Energy Efficiency* (Rep.). Available at www1.eere.energy.gov/buildings/publications/pdfs/alliances/commercial_building_energy_alliances_fact_sheet.pdf.

Elkington, J. (2004). Enter the triple bottom line. In Henriques, A. and Richardson, J. (Eds.) *The Triple Bottom Line: Does It All Add Up?* (pp. 1–16). London: Earthscan.

Energy Star (2017a). Energy Star overview. Available at www.energystar.gov/about/.

Energy Star (2017b). Building contest. Available at www.energystar.gov/index.cfm?c= healthcare.bus_healthcare_ny_presb_hospital.

Environmental Working Group, Public Affairs (2009). Toxic chemicals found in minority cord blood. Press release, December 2. Available at www.ewg.org/news/news-releases/2009/12/02/toxic-chemicals-found-minority-cord-blood#.WcUy6tOGPfZ.

Epstein, R. (1995). Communication between primary care physicians and consultants. *Archives of Family Medicine* 4(5): 403–9.

Fairfax County Board of Supervisors adopts green building policy (2008). Available at www.fairfaxcounty.gov/news/2008/030.htm.

Fast Facts Virginia Mason (2016). Available at www.virginiamason.org/workfiles/enviromason/enviromason_fastfacts.pdf.

Fauntleroy, G. (2007). Obesity leads to more hospital admissions, longer stays. *Health Behavior News Service*, December 11. Available at www.cfah.org/hbns/archives/getDocument.cfm?documentID=1631.

Federation of American Hospitals (2003). FTC/DOJ hearings on health care and competitive law and policy statement of the Federation of American Hospitals – hospital's non-profit status. Federal Trade Commission. Available at www.ftc.gov/sites/default/files/documents/public_events/health-care-competition-law-policy-hearings/completeagenda.pdf.

Federation of American Hospitals (2017). Available at https://fah.org/.

Ferracone, R. (2011). The role of environmental sustainability in executive compensation. *Forbes*, April.

Fiksel, J., McDaniel, J. and Mendenhall, C. (1999). Measuring progress towards sustainability principles, process, and best practices. *Greening of Industry Network Conference* (pp. 1–25). Available at www.eco-nomics.com/images/Sustainability%20Measurement%20GIN.pdf.

Flower, J. (1993). Getting paid to keep people healthy: two ways of integrating a healthcare system. *Healthcare Forum Journal* 36(2): 51–6.

Food & recipes: healthy eating (2009). *WebMD.com*, October 12. Available at www.webmd.com/food-recipes/tc/healthy-eating-overview.

FoodRoutes Network (2017). Buy fresh buy local. Available at http://foodroutes.org/buy-fresh-buy-local-program/.jsp.

Freund, F. (2009). *Business Model Concepts in Corporate Sustainability. Contexts*. Lüneburg, Germany: Leuphana Centre for Sustainability Management.

Frith, K. (2007). Is local more nutritious? It depends. *Center for Health and the Global Environment, Harvard Medical School*. Available at www.chgeharvard.org/sites/default/files/resources/local_nutrition.pdf.

Future of hospital care: how will we pay the bill? (2005). *A Hospital Payment Symposium*. Washington, DC: Federation of American Hospitals.

Galbraith, K. (2009, June 10). Study cites strong green job growth. *New York Times*, June 10. Available at http://green.blogs.nytimes.com/2009/06/10/study-cites-strong-green-job-growth.

Gallup, Inc. (2017). Drive employee engagement. Available at http://workplace.gallup.com/215921/optimize-employee-engagement.aspx.

Gershon, R.R.M., Pogorzelska, M., Qureshi, K.A., Stone, P.W., Canton, A.N., Samar, S.M. and Sherman, M. (2008). Home health care patients and safety hazards in the home: preliminary findings. *Agency for Healthcare Research and Quality*. Available at www.ahrq.gov/downloads/pub/advances2/vol. 1/Advances-Gershon_88.pdf.

Gidlason, D.S. (2017). Public health vs. the doctor's office. *Nutramed*. Available at www.nutramed.com/medicalcare/public_health.htm.

Global Environmental Management Initiative (2004). *Environment: Value to the Investor* (Mission Statement). Available at www.gemi.org/resources/GEMI%20Clear%20Advantage.pdf.

Global Green and Healthy Hospitals (2017). Available at http://greenhospitals.net.

Gold, S. (2011). How European nations run national health services. *Guardian*, May 11. Available at www.theguardian.com/healthcare-network/2011/may/11/european-healthcare-services-belgium-france-germany-sweden.

Gordon, E. (2012). Fast food's slow exit from hospitals. *Kaiser Health News*, April. Available at http://capsules.kaiserhealthnews.org/index.php/2012/04/fast-foods-slow-exit-from-hospitals.

Gow, P. (2012). *An Experience of "Yes": Independent Schools Begin to Explore and Exploit the Power of Design Thinking.* Washington, DC: National Association of Independent Schools. Available at www.nais.org/magazine/independent-school/spring-2012/an-experience-of-yes/.

GRACE Communications Foundation (2009). Why buy local? Available at www.sustainabletable.org/issues/whybuylocal/#fn8.

Green Building Council (2017). Available at www.usgbc.org.

Griffith, J.R. and White, K.R. (2007). *The Well-managed Healthcare Organization* (6th edn). Chicago, IL: Health Administration Press.

Hand, L. (2009). Employer health incentives. *Harvard School of Public Health.* Available at www.hsph.harvard.edu/news/magazine/winter09healthincentives/.

Health Care Spending in the United States and Selected OECD Countries (Rep.) (2011). April 12. Available at www.kff.org/health-costs/issue-brief/snapshots-health-care-spending-in-the-united-states-selected-oecd-countries/.

Health Care Without Harm (2017a). Available at www.hcwh.org/.

Health Care Without Harm (2017b). *The Campaign for Environmentally Responsible Health Care* (PowerPoint Presentation). Reston, VA. Available at http://infohouse.p2ric.org/ref/16/15197.pdf.

Healthier Hospitals Initiative (2017a). Available at http://healthierhospitals.org.

Healthier Hospitals Initiative (2017b). Advocate health care. Available at http://healthierhospitals.org/get-inspired/leadership-spotlight/advocate-health-care.

Health Resources & Services Administration (2017). Hill–Burton free and reduced-cost health care. U.S. Department of Health and Human Services. Available at www.hrsa.gov/get-health-care/affordable/hill-burton/index.html.

Highfield, R. (2006). The reality is that everything is made of chemicals. *Telegraph,* November 8. Available at www.telegraph.co.uk/news/uknews/1533519/The-reality-is-that-everything-is-made-of-chemicals.html.

How much of the Earth is covered with water? (2017). *Fairfax County Water Authority.* Available at www.fcwa.org/education/education.htm.

How much water is there on, in, and above the Earth? (1984). *US Geological Survey.* Available at http://ga.water.usgs.gov/edu/earthhowmuch.html.

Hyde, L. (2011). Confessions of a formal education enabler. *Museum Commons,* August 8. Available at http://museumcommons.blogspot.com/2011_08_01_archive.html.

Inova's Environmental Leadership (2017). Available at www.inova.org/about-inova/sustainability/history.jsp.

International Finance Corporation (2017). Overview. Available at www.ifc.org/wps/wcm/connect/Topics_Ext_Content/IFC_External_Corporate_Site/Sustainability-At-IFC.

Jarousse, L.A. (2012, January 1). Containing energy expenses. *Hospitals and Healthcare Networks,* January 1. Available at www.hhnmag.com/articles/5790-containing-energy-expenses.

Johnson, J. (2008). Rapporteur's report: services sector. *Journal of Safety Research* 39: 191–4.

Kaiser Permanente unveils sustainability scorecard for medical products (2010). *Environmental and Energy Management News,* May 4. Available at www.environmentalleader.com/2010/05/04/kaiser-permanente-launches-sustainability-scorecard-for-medical-products/.

Kanter, R.M. (2008). A financial turnaround requires culture change. *Harvard Business Review,* October 29. Available at http://blogs.hbr.org/kanter/2008/10/a-financial-turnaround-require.html.

Klein, E. (2012). Why an MRI costs $1,080 in America and $280 in France. *Washington Post,* March 3. Available at www.washingtonpost.com/blogs/ezra-klein/post/why-an-mri-costs-1080-in-america-and-280-in-france/2011/08/25/gIQAVHztoR_blog.html.

Krebs, A. (n.d.). Corporate agribusiness: monopolising subsistence. *Converge.* Available at www.converge.org.nz/pirm/corpag.htm.

Lacy, P. (2010). On the verge of a sustainability "tipping point." *GreenBiz*, June.

Lantz, P., Lichtenstein, R. and Pollack, H. (2007). Health policy approaches to population health: the limits of medicalization. *Health Affairs* 26(5): 1253–57.

Lean Enterprise Institute (2009). *What is Lean?* Cambridge, MA: Lean Enterprise Institute. Available at www.Lean.org/whatsLean.

Lewis, A. (2011). How my company hires for culture first, skills second. *Harvard Business Review*, January 26.

Linn, A. (2007). Corporations find business case for going green. *NBCNEWS.com*, April 18. Available at www.msnbc.msn.com/id/17969124/ns/business-going_green/t/corporations-find-business-case-going-green/#.T8RMcNVPujN.

Lipson, D.J. and Simon, S. (2010). Quality's new frontier: reducing hospitalizations and improving transitions in long-term care. *Mathematica Issue Brief* 7. Available at https://ideas.repec.org/p/mpr/mprres/307fb0f0db884e1699dd1045a1319c2d.html.

Lo, L. (2011). *Teamwork and Communication in Healthcare*. Edmonton, AB: Canadian Patient Safety Institute.

Manos, D. (2010). Experts name 9 ways to fix healthcare workforce shortage. *Health Care Finance News*, January 15. Available at www.healthcarefinancenews.com/news/experts-name-9-ways-fix-healthcare-workforce-shortage.

Martinez, S., Hand, M., Da Pra, M., Pollack, S., Ralston, K., Smith, T. and Newman, C. (2010). *Local Food Systems: Concepts, Impacts, and Issues* (Rep.). Washington, DC: U.S. Department of Agriculture. Available at www.ers.usda.gov/publications/pub-details/?pubid=46395.

Mathur, Y. (2009). Health care reform explained. *Business Today*. Available at www.businesstoday.org/magazine/its-always-christmas-washington/health-care-reform-explained.

McCain, M. (2011). *Ambulatory Care of the Future: Optimizing Health, Service, and Cost by Transforming the Care Delivery Model* (Rep.). The Chartis Group. Available at www.chartis.com/files/pdfs/Ambulatory_Care_of_the_Future.pdf.

McKinney, M. (2011). Outsourcing sees stimulus effect. *Modern Healthcare*, September.

McKinsey & Company (2008). How companies think about climate change: A McKinsey global survey. *McKinsey Quarterly*, February.

Medicaid.gov (2017a). Program of all-inclusive care for the elderly. Available at www.medicaid.gov/medicaid/ltss/pace/index.html.

Medicaid.gov (2017b). State resources. Available at www.npaonline.org/website/article.asp?id=203.

Medicine and Public Health Initiative (2012). Sustainability. Available at www.mphi.net/?p=303&lang=en-us.

Mikkonen, J. and Raphael, D. (2010). *Social Determinants of Health: The Canadian Facts*. Toronto, ON: York University School of Health Policy and Management. Available at www.unnaturalcauses.org/assets/uploads/file/The_Canadian_Facts.pdf.

Minnesota Life & Securian (2008). *The New Buzz in Health: Wellness Programs Offer Savings for Employers, a New Revenue Source for Producers* (Rep.). Available at www.lifebenefits.com/lb/pdfs/F62382-20%20Get%20More%2012.pdf.

Mufson, S. and Yang, J.L. (2011). A quarter of US nuclear plants not reporting equipment defects, report finds. *Washington Post*, March 24.

Muro, M., Rothwell, J. and Saha, D. (2011). Sizing the clean economy: a national and regional green jobs assessment. *Brookings Report*, July 13. Available at www.brookings.edu/research/sizing-the-clean-economy-a-national-and-regional-green-jobs-assessment/.

Myslymi, E. (2016). U.S. News & World Report releases 2017 best colleges rankings. September. Available at www.usnews.com/info/blogs/press-room/articles/2016-09-13/us-news-releases-2017-best-colleges-rankings (accessed August 11, 2017).

National Cancer Institute (2016). President's cancer panel: annual reports. Available at http://deainfo.nci.nih.gov/ADVISORY/pcp/annualReports/index.htm.

National Prevention Council (2011). *National Prevention Strategy: America's Plan for Better Health and Wellness* (Rep.). Rockville, MD: National Prevention, Health Promotion and Public Health Council. Available at www.surgeongeneral.gov/priorities/prevention/strategy/report.pdf.

Newby-Clark, I. (2009). Creatures of habit. *Psychology Today*, July 17. Available at www.psychologytoday.com/blog/creatures-habit/200907/we-are-creatures-habit.

NICE (2008a). *Guidance on the Promotion and Creation of Physical Environments that Support Increased Levels of Physical Activity.* London: National Institute for Health and Clinical Excellence. Available at www.nice.org.uk/guidance/ph8.

NICE (2008b). *Physical Activity and the Environment.* London: National Institute for Health and Clinical Excellence. Available at www.nice.org.uk/guidance/ph8.

NIH (2002). "I swear by Apollo physician [...]" Greek medicine from the Gods to Galen. *National Library of Medicine.* Available at www.nlm.nih.gov/hmd/greek/greek_oath.html.

NNEC (2017). *The Energy–Water Nexus.* New York: Network for New Energy Choices. Available at www.newenergychoices.org/index.php?page=ew_intro&sd=ew.

Northbridge Environmental Management Consultants (2003). *Analyzing the Cost of Obtaining LEED Certification* (Rep.). Westford, MA: Northbridge for the American Chemistry Council. Available at https://greenbuildingsolutions.org/wp-content/uploads/2016/05/LEED-Cost-Analysis-Report.pdf.

O'Brien-Pallas, L., Griffin, P., Shamian, J., Buchan, J., Duffield, C., Hughes, F. and Stone, P.W. (2006). The impact of nurse turnover on patient, nurse, and system outcomes: a pilot study and focus for a multicenter international study. *Policy Politics Nursing Practice* 7(3): 169–79.

Odell, A.M. (2007). Working for the Earth: green companies and green jobs attract employees. *GreenBiz*, October 16. Available at www.greenbiz.com/news/2007/10/16/working-earth-green-companies-and-green-jobs-attract-employees.

Oglethorpe, D. (2008). Local food: miles better? *World Watch* 22(3): 12–15.

Ornish, D. (2009). Mostly plants. *American Journal of Cardiology* 104(7): 957–8.

Ornish, D. (2012). Holy Cow! What's good for you is good for our planet. *Archives of Internal Medicine*: E9–E10.

Paris, L. (2011). Pew finds serious gaps in oversight of US drug safety. *Pew Health Group Safety*, July. Available at www.pewhealth.org/news-room/press-releases/pew-finds-serious-gaps-in-oversight-of-us-drug-safety-85899367931.

Parker-Pope, T. (2008). Boosting health with local food. *New York Times Well Blogs*, June 6. Available at http://well.blogs.nytimes.com/2008/06/06/boosting-health-with-local-food/.

Partners HealthCare System (2009). Interview. Massachusetts.

Pho, K. (2011). Primary care doctors and specialists need to better communicate. *KevinMD.com*, January. Available at www.kevinmd.com/blog/2011/01/primary-care-doctors-specialists-communicate.html.

Physicians for Social Responsibility (2013). Hazardous chemicals in health care. Available at www.psr.org/resources/hazardous-chemicals-in-health.html.

Poladian, C. (2012). Asthma rates continue to soar in the US. *Medical Daily*, May 17. Available at www.medicaldaily.com/asthma-rates-continue-soar-us-240475.

Practice Greenhealth (2011). Greening the OR: fluid management systems in the OR. Available at http://practicegreenhealth.org/webinars/greening-or-fluid-management-systems-or.

Practice Greenhealth (2012a). *Sample Job Description: Healthcare Sustainability Director.* Reston, VA: Practice Greenhealth.

Practice Greenhealth (2012b). Practice Greenhealth 2012 Environmental Excellence Awards. Available at http://practicegreenhealth.org/awards/.

Practice Greenhealth (2017a). Available at http://practicegreenhealth.org.

Practice Greenhealth (2017b). How health care uses energy. Available at https://practice greenhealth.org/topics/energy-water-and-climate/energy.

Report of the World Commission on Environment and Development: Our Common Future (p. 43, Rep.) (1987). New York: United Nations.

Republic of Ireland Department of Health and Children (2005). *Waste Management in Hospitals* (Rep.). Available at http://audgen.gov.ie/documents/vfmreports/49_Waste_Management_in_Hospitals.pdf.

Rich, C., Singleton, J. and Wadhwa, S. (2011). Teaching sustainability as part of healthcare management studies: challenges, best practices and case studies. *CleanMed*.

Rich, C., Wadhwa, S. and Singleton, J. (2012). How to teach environmental sustainability in healthcare management education. In *Teaching Sustainability*. Allendale, MI: Grand Valley State University.

Ross, A. (2010). *Planning, Transport, and Health Inequalities: A Recent History and Future Progress* (Rep.). UK Local Government Association. Available at www.idea.gov.uk/idk/core/page.do?pageId=23454074.

Roylance, F. (2007). Manufacturers may be the only ones cleaning up. *Los Angeles Times*, February 11. Available at http://articles.latimes.com/2007/feb/11/news/adna-germs11/2.

Sabrina, A. (2010, May 8). America's obesity epidemic to cost $550 billion by 2030: study. *International Business Times*, May 8. Available at www.ibtimes.com/articles/338553/20120508/america-s-obesity-epidemic-hangs-550-billion.htm.

Satin, D. and Miles, J. (2009). Performance-based bundled payments: benefits and burdens of this pay-for-performance strategy (Rep.). *Minnesota Medicine*, October. Available at www.minnesotamedicine.com/PastIssues/October2009/SpecialReportOct2009/tabid/3209/Default.aspx.

Sg2: Demand for outpatient services to increase by nearly 22% over the next decade (2010). *News Medical*, January 6. Available at www.news-medical.net/news/20100106/Sg2-Demand-for-outpatient-services-to-increase-by-nearly-2225-over-the-next-decade.aspx.

Shontell, A. (2011). 11 Unusual ways Steve Jobs made Apple the world's most admired tech company. *Business Insider*, October 6. Available at www.businessinsider.com/steve-jobs-apple.

Singleton, J. (1992). Preserving the environment can bring hospitals savings, improved stature. Presented at Modern Healthcare Weekly Business News: Our Environment: A Healthcare Commitment. Arlington, VA, March 8–9.

Smith, C. (2007). *Managing Pharmaceutical Waste: A 10-step Blueprint for Healthcare Facilities*. Houston, TX: PharmEcology Associates, LLC.

Studer, Q. (2003). *Hardwiring Excellence: Purpose, Worthwhile Work, Making a Difference*. Gulf Breeze, FL: Fire Star Publishing.

Taylor, D. (2002). What price for-profit hospitals? *Canadian Medical Association Journal* 166(11): 1418–19.

Terry, K. (2009). Hospital EHRs don't make paper go away. *CBS News*, July 29. Available at www.cbsnews.com/8301-505123_162-43840588/hospital-ehrs-dont-make-paper-go-away/.

The economic benefits of sustainable design (n.d.). Available at www1.eere.energy.gov/femp/pdfs/buscase_section2.pdf.

The Food Trust (2017). Available at http://thefoodtrust.org/food-access/publications.

The Global Importance of More Sustainable Products in The Global Health Care Industry (n.d.). A study commissioned by Johnson & Johnson.

Thomas (2011). China's health reforms progress. *International Insurance News*, March 11. Available at www.globalsurance.com/blog/chinas-healthcare-reforms-progress-324520. html.

Towers Watson & National Business Group on Health (2010). *Raising the Bar on Health Care*. Available at www.willistowerswatson.com/DownloadMedia.aspx?media=%7B4A0241 10-2738-42EE-8F14-7EF06F4B839D%7D.

Transportation is a public health issue; DOT doing its part to keep kids moving (2010). United States Department of Transportation, July 16. Available at http://usdotblog. typepad.com/secretarysblog/2010/07/dot-doing-its-part-to-keep-kids-moving.html.

Tsai, T. (2010). Second chance for health reform in Columbia. *The Lancet* 375(9709): 109–10.

Turley, M., Porter, C., Garrido, T., Gerwig, K., Young, S., Radler, L. and Shaber, R. (2011). Use of electronic health records can improve the health care industry's environmental footprint. *Health Affairs* 30(5): 938–46.

Umbdenstock, R. (2011). Hospital environmental sustainability: video intro. American Hospital Association. Available at www.hospitalsustainability.org.

U.S. Bureau of Labor Statistics (2010). *The 2010 President's Budget for the Bureau of Labor Statistics* (Rep.). Available at www.bls.gov/bls/budget2010.htm.

U.S. Department of Health and Human Services. (2008). *Prevention makes common "cents"* (Rep.). Washington, DC. Retrieved from http://aspe.hhs.gov/health/prevention.

U.S. Department of Transportation (2011). *Pipelines and Hazardous Materials*. Washington, DC.

U.S. Environmental Protection Agency (1970). *Clean Air Act*. Washington, DC. Available at www.epa.gov/air/caa/.

U.S. Environmental Protection Agency (2004a). *Code of Federal Regulations: Title 40 Protection of Environment, Section 311.1*. Washington, DC.

U.S. Environmental Protection Agency (2004b). *Leaders in Healthcare Tap the Power of Superior Energy Management*. Washington, DC. Available at www.energystar.gov/ia/business/ healthcare/factsheet_0804.pdf?9b81-81ae.

U.S. Environmental Protection Agency (2010a). *Essential Principles for Reform of Chemicals Management Legislation*. Washington, DC. Available at www.epa.gov/assessing-and-managing-chemicals-under-tsca/essential-principles-reform-chemicals-management-0.

U.S. Environmental Protection Agency (2010b). *Green Building*. Washington, DC. Available at www.epa.gov/greenbuilding/pubs/about.htm.

U.S. Environmental Protection Agency (2011). *Benefits and Costs of the Clean Air Act: Second Prospective Study – 1990 to 2020*. Washington, DC. Available at www.epa.gov/clean-air-act-overview/benefits-and-costs-clean-air-act-1990-2020-second-prospective-study.

U.S. Environmental Protection Agency (2012). *Medical Waste*. Washington, DC. Available at www.epa.gov/rcra/medical-waste.

U.S. Environmental Protection Agency. (2017). Retrieved from www.epa.gov.

U.S. Food and Drug Administration (2010). *Triclosan: What Consumers Should Know*. Washington, DC. Available at www.fda.gov/forconsumers/consumerupdates/ucm205999. htm.

U.S. Occupational Safety and Health Administration (2011). *Hazardous Waste*. Washington, DC.

University of California Santa Cruz (2017). Office of Planning & Budget. Available at https://planning.ucsc.edu/.

Walsh, D. (2012). Taking sustainability seriously. *Green at Work*, January 27. Available at http://greenatwork.com/blog/2012/01/27/taking-sustainability-seriously/.

Warde, A. (2017). Going green to stay in the black. University of California. Available at www.uci.edu/features/2011/02/feature_marriott_110214.php.

Waste Prevention and Recycling Posters and Stickers (2006). The California Integrated Waste Management Board. Available at www.calrecycle.ca.gov/publications/Documents/BizWaste/44105012.pdf.

Watters, C., Sorensen, J., Fiala, A. and Wismer, W. (2003). Exploring patient satisfaction with foodservice through focus groups and meal rounds. *Journal of the American Dietetic Association* 103(10): 1347–49.

White, G. (2009). Introduction to green leasing. *Green Real Estate Law Journal*, February 6. Available at www.greenrealestatelaw.com/2009/02/introduction-to-green-leasing.

Whitlock, J.L. (2003). *Strategic Thinking, Planning, and Doing: How to Reunite Leadership and Management to Connect Vision with Action* (Rep.). Washington, DC: George Washington University Center for Excellence in Municipal Management.

Wickel, R. (2012). From class to farm: students grow understanding of food production. *The Pendulum: Elon University's Student News Organization*, May 1. Available at http://elonpendulum.com/2012/05/from-class-to-farm-students-grow-understanding-of-food-production.

World Health Organization & Health Care Without Harm (2009). *Healthy Hospitals – Healthy Planet – Healthy People*. Geneva, Switzerland: World Health Organization. Available at www.who.int/globalchange/publications/climatefootprint_report.pdf.

World Wildlife Fund (2011). Hospital waste factsheet. Available at www.wwfpak.org/factsheets_hwf.php.

INDEX

Page numbers in *italics* denote tables, those in **bold** denote figures.